Standard of Ministry of Water Resources of
the People's Republic of China

SL 25—2006
Replace SL 25—91

Design Specification for Stone Masonry Dam

Drafted by:
Water Resources Department of Guizhou Province

Translated by:
Water Resources Department of Guizhou Province

China Water & Power Press
Beijing 2017

图书在版编目（CIP）数据

浆砌石坝设计规范：SL25-2005：Design Specification for Stone Masonry Dam SL25-2005：英文 / 中华人民共和国水利部发布. -- 北京：中国水利水电出版社，2017.6
ISBN 978-7-5170-5626-3

Ⅰ. ①浆… Ⅱ. ①中… Ⅲ. ①浆砌石－砌石坝－设计规范－中国－英文 Ⅳ. ①TV641.3-65

中国版本图书馆CIP数据核字(2017)第172749号

书　　名	Design Specification for Stone Masonry Dam SL 25—2006
作　　者	中华人民共和国水利部　发布
出版发行	中国水利水电出版社 （北京市海淀区玉渊潭南路1号D座　100038） 网址：www.waterpub.com.cn E-mail：sales@waterpub.com.cn 电话：(010) 68367658（营销中心）
经　　售	北京科水图书销售中心（零售） 电话：(010) 88383994、63202643、68545874 全国各地新华书店和相关出版物销售网点
排　　版	中国水利水电出版社微机排版中心
印　　刷	北京瑞斯通印务发展有限公司
规　　格	140mm×203mm　32开本　3.625印张　127千字
版　　次	2017年6月第1版　2017年6月第1次印刷
定　　价	**260.00元**

凡购买我社图书，如有缺页、倒页、脱页的，本社营销中心负责调换

版权所有·侵权必究

Introduction to English Version

China Water Conservancy and Hydropower Investigation and Design Association (CWHIDA) is a non-profit, nationalwide and self-regulatory organization under the administration of Ministry of Water Resources of the People's Republic of China, which has 390 members. In order to satisfy the increasing demand of international water conservancy and hydropower engineering construction, CWHIDA has formulated the *Translation Plan of Technical Standards 2015—2017* and organized its members to translate some technical standards from Chinese into English.

The translation of SL 25—2006 *Design Specification for Stone Masonry Dam* from Chinese into English was organized by CWHIDA in accordance with due procedures and regulations applicable in the country. The translation was undertaken by Water Resources Department of Guizhou Province.

The English version of this standard is identical in force to its Chinese original SL 25—2006 *Design Specification for Stone Masonry Dam* which includes the table of contents and main body of the Chinese original SL 25—2006. In case of any discrepancy, the Chinese original shall prevail.

Translation task force includes ZHANG Ming, LIU Qi, YANG Xiaochun, ZHANG Yanhong, YANG Chaohui, LI Jin, JIN Feng, AN Xuehui, ZHOU Hu, ZHAO Yun, ZENG Xinbo, CHEN Changjiu, HUANG Miansong, LI Fengliang, JIANG Maoxi, XIONG Jie, and SHEN Ruozhu.

The English version of this specification was reviewed by

ZHU Dangsheng, LEI Xingshun, DU Leigong, ZHANG Guoxin, CHEN Shangfa, CHEN Liqiu, LUO Sujuan, HE Qirong, ZHOU Jianguo, LAN Guangyu, WEN Minggong, and LIU Jiahui.

China Water Conservancy and Hydropower Investigation and Design Association

Foreword

It has been more than 10 years since SL 25—91 *Design Specification for Stone Masonry Dam* (hereinafter referred to as the original specification) was initially issued in 1991. According to *Notice on the Plan for the Formulation and Revision of Technical Standards and Chief Editing Organization of Technical Standards for Investigation and Design of Water Resources and Hydropower* in 2002 (File No. SGJ 2002 [15]), issued by China Renewable Energy Engineering Institute of Ministry of Water Resources, and SL 1—2002 *Specification for the Drafting of Technical Standards of Water Resources*, the original specification has been amended and named SL 25—2006 *Design specification for Stone Masonry Dam* (hereinafter referred to as this specification). Masonry dams in this specification include mortar masonry dams or concrete masonry dams exclusive of dry-laid masonry dams.

The main contents of amendments and supplements to the original specification are as follows:

—A new chapter, "Terms and Symbols", is added following "General Provisions", increasing the number of chapters from the original 9 to the present 10.

—A load combination of reservoir design flood level and the design temperature load of normal temperature rise is added to basic combination of loads. A specified period for calculation of sediment deposition is included in Section 4.1 of this specification (see Annex D).

—Section 4.2 "Dam Shape Design" of the original specification is changed to Section 5.2 "Dam Structure" of this specification, in which the structures of masonry gravity dams are stipulated. As for the stability against deep-seated sliding mentioned in the original Clause 4.3.4 (Clause 5.3.3 in this specification), Annex E is added and safety factor as well as its calculation formula against deep-seated sliding of masonry gravity dams are specified.

—Important modifications are made to Table 5.2.5-1 in the original specification (see Table A.0.7 in this specification), adding two types of stone masonry, i.e. 70% rubble plus 30% ashlar masonry and 30% rubble plus 70% ashlar masonry, which is more economically viable.

—Chapter 9 "Observation Design" in the original specification is changed to Chapter 10 "Safety Monitoring Design" in this specification, in which the scope of monitoring is stipulated, the principles for safety monitoring design are supplemented, and the requirements for main facility layout are added.

—Annex Ⅲ and Annex Ⅴ in the original specification are removed. Annexes Ⅰ, Ⅱ, Ⅳ and Ⅵ are changed, respectively, to Annexes A, C, G, and B.

This specification replaces SL 25—91 *Design Specification for Stone Masonry Dam*.

This specification is approved by Ministry of Water Resources of the People's Republic of China.

This specification is interpreted by General Institute of Water Resources and Hydropower Planning and Design, Ministry of Water Resources.

This specification is chiefly drafted by Department of Water Resources of Guizhou Province.

This specification is jointly drafted by Hunan Hydro and

Power Design Institute, Fujian Provincial Investigation of Design and Research Institute of Water Conservancy and Hydropower, Guizhou Survey and Design Research Institute for Water Resources and Hydropower, Zunyi Survey and Design Institute of Water Conservancy and Hydropower.

Chief drafters are LI Zhanmei, WANG Liangzhi, GAO Shibao, HE Qirong, and YANG Weizhong.

Leading experts of the technical review meeting are SHEN Fengsheng and LEI Xingshun.

Format examiner is CHEN Dengyi.

Contents

Introduction to English Version

Foreword

1 General Provisions ·· 1
2 Terms and Symbols ··· 4
   2.1 Terms ··· 4
   2.2 Symbols ·· 5
3 Dam Materials and Design Indexes of Stone Masonry ··· 7
   3.1 Dam Materials ··· 7
   3.2 Design Indexes of Stone Masonry ··· 9
4 Loads and Load Combinations ··· 11
   4.1 Loads ·· 11
   4.2 Load Combinations ··· 14
5 Stone Masonry Gravity Dam ··· 18
   5.1 Dam Layout ··· 18
   5.2 Dam Structure ··· 19
   5.3 Calculation of Dam Sliding Stability ··· 20
   5.4 Calculation of Dam Stresses ··· 22
   5.5 Dam Temperature Control and Crack Prevention ················· 25
6 Stone Masonry Arch Dam ·· 26
   6.1 Dam Site, Dam Axis and Dam Layout ····································· 26
   6.2 Dam Stress Analysis ·· 27
   6.3 Abutment Stability Analysis ··· 29
   6.4 Temperature Control ·· 31
7 Dam Seepage Control ·· 33
   7.1 General Provisions ·· 33
   7.2 Impervious Concrete Face Slab and Core ································ 33

7.3	Seepage Control of Dam Body	35
7.4	Transverse Joints, Water Stop and Drainage	35
8	Treatment of Dam Foundation	37
9	Dam Structure	40
9.1	Crest Layout and Transportation	40
9.2	Galleries and Openings	41
9.3	Joints, Drainage and Foundation Bedding	42
10	Safety Monitoring Design	44
10.1	General Principles	44
10.2	Monitoring Items and Monitoring Facility Layout	46
Annex A	Main Mechanical Indexes of Stone Masonry	50
Annex B	Test Methods of Deformation (Elastic) Modulus and Compressive Strength of Stone Masonry	65
Annex C	Calculation Formulas of Loads	69
Annex D	Time Limitation for Calculations of Sedimentation in Front of Dam	89
Annex E	Safety Factors of Deep Sliding Stability and the Formulas for Stone Masonry Material Mechanics Dam Foundation	90
Annex F	Stress Calculations of Stone Masonry Gravity Dam with the Gravity Method, Considering the Distinct Elastic Moduli of Face Slab and Stone Masonry	94
Annex G	Calculation of Stresses of Gravity Block and Thrust Block with the Material Mechanics Method	101
Explanation of Wording		105

1 General Provisions

1.0.1 To keep pace with development of stone masonry dam construction and standardize the design of stone masonry dams, SL 25—91 *Design Specification for Stone Masonry Dam* is revised to this specification so that the stone masonry dam design could be safe and applicable, economically viable, technology updated and quality ensured.

1.0.2 This specification is applicable to the design of Class 2 and Class 3 masonry dams for large and medium water resources and hydropower projects or Class 4 and Class 5 masonry dams over 50 m high. It may also be used as a reference for the design of other masonry dams.

For stone masonry dams over 100 m high, special studies shall be conducted.

1.0.3 Stone masonry dams are classified into low dams, medium-height dams and high dams by the dam height. Low dams are those below 30 m high, medium dams are those 30 to 70 m high, and high dams are those over 70 m high.

1.0.4 Stone masonry arch dams are classified into thin arch dams, medium-thickness arch dams and thick arch dams (or referred to as gravity arch dams) according to the ratio of thickness to height. Thin arch dams are those with thickness-height ratio less than 0.2, medium ones 0.2 to 0.35, and thick ones greater than 0.35.

1.0.5 In the design of stone masonry dams, the following issues shall be emphasized and studied:

    1 The basic data on the dam site and its vicinity concerning river planning, comprehensive function requirements, hydrology, meteorology, topography, geology, earthquake, building materials, conditions for construction and operation, etc.

2 Dam shape, dam site and dam axis to be reasonably selected, dam structure to be simplified, and the scheme to be optimized as far as possible through comprehensive study on the relationship between the dam layout and other structures in conjunction with the general layout of project.

3 The treatment of dam foundation and the seepage control of dam body.

4 The flood release, energy dissipation and scouring protection.

5 The river diversion and the safety for flood control during construction.

6 The proper use of building materials, construction modes and construction technologies in consideration of local conditions. New materials, new workmanship and new technologies shall be used actively and prudently on the basis of experience from established practice and scientific experiments.

7 The measures to reduce project cost and shorten the construction period.

In addition, the similarities and differences in the designs between stone masonry dams and concrete dams of the same type shall be studied. Enough attention shall be paid to material tests and structural calculations and analyses of stone masonry dams, so as to gradually explore and apply designs and calculation methods reflecting structural characteristics of stone masonry dams.

1.0.6 Provisions stipulated not only in this specification but also in the prevailing national standards shall be complied with for the design of masonry dams.

1.0.7 The standards cited in this specification mainly include:

GB 50287—99 *Code for Water Resources and Hydropower Engineering Geological Investigation*

SL 252—2000 *Standard for Classification and Flood Control of Water Resources and Hydroelectric Project*

SL 203—97 *Specifications for Seismic Design of Hydraulic Structures*

SL/T 191—96 *Design Code for Hydraulic Concrete Structures*

SL 211—98 *Code for Design of Hydraulic Structures against Ice and Freezing Action*

SL 319—2005 *Design Specification for Concrete Gravity Dams*

SL 282—2003 *Design Specification for Concrete Arch Dams*

SD 105—82 *Test Code for Hydraulic Concrete*

SL 264—2001 *Specifications for Rock Tests in Water Conservancy and Hydroelectric Engineering*

SDL 336—89 *Technical Specification for Concrete Dam Safety Monitoring*

2 Terms and Symbols

2.1 Terms

2.1.1 Stone masonry dam

Masonry dams constructed of stone units bonded by cement mortar or concrete with one - or two - graded aggregates. The main types are stone masonry gravity dams and stone masonry arch dams.

2.1.2 Dam height

The vertical distance from the dam crest to the lowest point (exclusive of the local deep trenches, wells or holes) of the excavated dam base.

2.1.3 Limit equilibrium method

An analysis method, based on limit equilibrium principle, to calculate safety factor of stability against sliding along slip plane of rock mass by assuming the rock mass as rigid body.

2.1.4 Trial - load method

An analysis method to determine the load distribution between the arches and cantilevers, by visualizing the arch dam consisting of a system of a series of horizontal arches or a system of a series of vertical cantilevers, under the condition that equal arch and cantilever deflections at all intersection points are produced by the load distribution.

2.1.5 Gravity block

Gravity structures that bear the arch abutment thrust by gravity action.

2.1.6 Thrust block

A structure between arch dam and bedrock to transfer thrust

of arch abutment to stable bedrock.

2.2 Symbols

2.2.1 Loads

P_{sk}—Silt pressure;

p_{sk}—Intensity of silt pressure;

P_{wk}—Wave pressure;

F_{hk}—Impact ice pressure by ice striking on dam face;

P_x—Horizontal component of resultant centrifugal force upon the reverse curve on the downstream face of spillway;

P_y—Vertical component of resultant centrifugal force upon the reverse curve on the downstream face of spillway;

α, α_1, α_2—Coefficients of uplift pressure on dam base;

T_m—Mean temperature change on a cross section;

T_d—Equivalent linear temperature change.

2.2.2 Material properties

γ_w—Unit weight of water (or water containing sediments);

ρ_w—Water density;

γ_{sd}—Dry unit weight of sediment;

γ_{sb}—Submerged weight of sediment;

ϕ_s—Internal friction angle of sediment;

R—Compressive strength of stone;

R_d—Dry compressive strength of stone;

R_s—Saturated compressive strength of stone;

γ_c—Unit weight of concrete;

E_c—Elastic modulus of concrete;

γ_b—Unit weight of stone masonry;

α—Linear expansion coefficient of stone masonry;

ρ_d—Dry density of stone masonry;

E_0—Deformation modulus of stone masonry;

E_e—Elastic modulus of stone masonry;

μ—Poisson's ratio of stone masonry;

λ_b—Thermal conductivity of stone masonry;

C_b—Specific heat of stone masonry;

a_b—Thermal diffusivity of stone masonry;

f_{cc}—Compressive strength of stone masonry;

f_t—Tensile strength of stone masonry;

f_f—Bending strength of stone masonry;

f'—Friction coefficient for shear-friction on slip plane;

c'—Cohesion for shear-friction on slip plane;

f—Friction coefficient for sliding-friction on slip plane.

2.2.3 Calculation indexes

K'—Safety factor of sliding stability calculated by shear-friction formula;

K—Safety factor of sliding stability calculated by sliding-friction formula.

3 Dam Materials and Design Indexes of Stone Masonry

3.1 Dam Materials

3.1.1 Stones used for masonry shall meet the following stipulations:

1 The stones shall be fresh, intact, hard, and free of spalls and cracks.

2 The stones may be classified by shape into three types, i.e., rubble, ashlar and coarse stone.

Rubble: irregular in size and shape. Single stone should be more than 25 kg in mass, and should not less than 20 cm in central thickness or local thickness.

Ashlar: with approximately hexahedron shape, nearly parallel and smooth top and bottom surfaces, and no sharp corners and thin edges. Its thickness should be greater than 20 cm.

The maximum length (length, width, and height) of rubbles and ashlars should not be greater than 100 cm.

Coarse stone: with clear edges and corners and six approximately flat surfaces. The maximum height difference on a surface should not be greater than 3% of its length. Its length should be greater than 50 cm and both its width and height should not be less than 25 cm.

3 The compressive strength of stones is classified into six grades, i.e., the grades of $\geqslant 100$ MPa, 80 MPa, 60 MPa, 50 MPa, 40 MPa, and 30 MPa, according to the saturated compressive strength.

4 Tests for physical and mechanical properties of stone

shall be conducted before the stone is put into use. For medium and small projects, the corresponding property values may be selected from Table A.0.1 in Annex A of this sepcification if tests cannot be conducted.

3.1.2 Aggregates used in stone masonry shall conform to the following stipulations:

1 The quality of aggregates shall meet the requirements specified in SD 105—82 *Test Code for Hydraulic Concrete*.

2 Fine aggregates include natural sand and crushed sand. Crushed sand shall be free of particles from soft rock and weathered rock. The maximum grain size of natural sand or crushed sand should be less than 5 mm.

3 Coarse aggregates (gravel, crushed aggregates) should be graded by diameter as: one grade that is 5 - 20 mm when the maximum diameter is 20 mm; two grades that are 5 - 20 mm and 20 - 40 mm when the maximum diameter is 40 mm.

3.1.3 Cementitious materials

1 Cementitious materials for stone masonry are mainly cement mortar and concrete with one - or two - graded aggregates.

2 The strength grades or specified strengths of cementitious materials adopted are as follows:

The strength grade is for cement and such three grades as 32.5, 42.5 and 52.5 are commonly used.

The specified strength is for cement mortar and four grades, i.e., 5.0 MPa, 7.5 MPa, 10.0 MPa, and 12.5 MPa, are commonly used.

The specified strength is for concrete and three grades, i.e., 10.0 MPa, 15.0 MPa, and 20.0 MPa, are commonly used.

Note: The specified strength here means the compressive

strength of cubic specimen of 15 cm×15 cm×15 cm tested at the age of 90 or 180 days with an assurance ratio of 80%.

3 The mix proportion of cementitious materials shall meet the requirements for the design strength of stone masonry.

4 Additives and admixtures should be used in cementitious material. The optimal amount shall be determined by tests.

3.2 Design Indexes of Stone Masonry

3.2.1 The design density of stone masonry, ρ_d, should be chosen according to the type of stone masonry in the following ranges:

For rubble masonry: $\rho_d = 2100 - 2350$ kg/m³.

For ashlar masonry: $\rho_d = 2200 - 2400$ kg/m³.

For coarse stone masonry: $\rho_d = 2300 - 2500$ kg/m³.

3.2.2 The linear expansion coefficient of stone masonry, α, may be chosen from $6 \times 10^{-6}/°C$ to $8 \times 10^{-6}/°C$, and if necessary may be determined by tests according to the method specified in SD 105—82.

3.2.3 The deformation parameters of stone masonry shall be selected based on the following stipulations:

1 Deformation modulus, E_0, and elastic modulus, E_e, of stone masonry should be determined by tests according to the method specified in Annex B of this specification. For projects for which tests cannot be conducted, the values may be selected from Table A.0.2 of Annex A in this specification.

2 Poisson ratio, μ, of stone masonry should be 0.2 - 0.25.

3.2.4 For Class 2 structures, ultimate compressive strength of stone masonry, f_{cc}, shall be determined by tests according to the method specified in Annex B of this specification. For Class 3 structures, it may be selected from Table A.0.3 of Annex A in this specification if tests cannot be conducted.

3.2.5 For Class 2 structures, tensile strength of stone masonry, f_t, shall be determined through tensile tests according to the method for fully-graded concrete specified in SD 105—82 or through in-situ measurements, while the laying method for specimens shall be the same as the actual construction method. For Class 3 structures, it may be selected from Table A.0.4 of Annex A in this specification.

3.2.6 In the preliminary design stage, the parameters for the sliding stability calculation of stone masonry dams, including the stability against sliding along the interface between bedding concrete and bedrock, the interface between stone masonry and bedding concrete, and along sections in the stone masonry, shall be determined by in-situ tests for Class 2 structures. For Class 3 structures, the parameters may be selected from Tables A.0.5 and A.0.6 in this specification according to the characteristics of bedrocks, strength of stone masonry, type and specified strength of cementitious materials, if tests cannot be conducted.

4 Loads and Load Combinations

4.1 Loads

4.1.1 Loads acting on stone masonry dams shall include dead weight, hydrostatic pressure, uplift pressure (or seepage pressure), silt pressure, wave pressure, ice pressure, hydrodynamic pressure, temperature load, seismic load, and other possible loads.

4.1.2 Dead weight: mainly covers the weight of dam body and the weight of permanent equipment. The unit weight of stone masonry shall be determined by tests, and it may be chosen according to Article 3.2.1 of this specification when test data are unavailable.

4.1.3 Hydrostatic pressure: upstream hydrostatic pressure shall be determined in accordance with the reservoir water level which is subject to the function of reservoir and load combinations, and the downstream hydrostatic pressure shall be determined by calculation according to Annex C.1 of this specification with respect to unfavorable tail water level. The unit weight of water should be 9.81 kN/m^3, and for sediment-laden rivers it shall be determined in accordance with specific conditions.

4.1.4 Uplift pressure: in the stability analysis and stress analysis of stone masonry gravity dams, and the foundation and abutments stability analysis of stone masonry arch dam, the uplift pressure or seepage pressure acting perpendicularly to the entire calculated section area shall be considered and calculated in accordance with Annex C.2 of this specification. In the stress analysis of stone masonry arch dam body, the effect of uplift pressure should be considered (uplift pressure may be neglected for thin arch dams).

4.1.5 Silt pressure: the sediment depth in front of dam shall be determined in accordance with the hydrological and sediment characteristics of the river, project general layout, reservoir operation conditions, computation of deposition and scouring, etc. For sediment-laden rivers, a special study shall be conducted. The period for computing deposition and scouring should conform to Annex D. Silt pressure shall be calculated in accordance with Annex C.3.

4.1.6 Wave pressure: shall be calculated in accordance with the wave elements (wave height and wave length). For reservoirs located in valley of mountainous area, it shall be determined in accordance with Annex C.4.

Different wind velocities should be used for different load combinations. The maximum annual wind velocity in a 50-year return period can be used for the usual combinations and the average of the long-term maximum annual wind velocity may be used for the unusual combinations.

4.1.7 Ice pressure: in severe cold regions where thick ice cover forms on the surface of reservoir, the effect of ice pressure shall be considered. The ice pressure shall include both static ice pressure and dynimic ice pressure, and should be determined in accordance with Annex C.5.

4.1.8 Hydrodynamic pressure: for flood discharging over the dam, the hydrodynamic pressure upon the reverse curve on the downstream face of spillway shall be considered. The calculation should be performed in accordance with Annex C.6. The pulsating pressure and negative pressure on overflow surface may be neglected.

4.1.9 Temperature load: the cases with respect to the normal design temperature drop and rise shall be considered respectively for stone masonry arch dams. The thermal load can be determined in accordance with the difference between dam temperature during operation and closure temperature. Temperature loads may be neglected for stone masonry gravity dams.

4.1.10 For defining temperature load, closure temperature field, average annual temperature field, and variable temperature field caused by the variation of surface temperature, shall be determined in accordance with Annex C. 7 of this specitication and the factors including environmental temperature near the dam site, reservoir temperature, solar radiation, dam foundation temperature, dam thickness, thermal properties of masonry materials, etc.

4.1.11 When the ratio is $d/R \leqslant 0.5$, in which d denotes the thickness of arch and R denotes the radius of horizontal arch, the influence of curvature of dam face may be negligible and the temperature field of arch dam may be calculated by regarding the arch as a slab.

The temperature may be divided into three parts according to its distribution through the dam thickness (Figure 4.1.11): (Ⅰ) average temperature variation through the section, (Ⅱ) equivalent linear temperature difference, and (Ⅲ) non-linear variation of temperature difference. Only parts (Ⅰ) and (Ⅱ) may be considered when calculating temperature load.

4.1.12 Seismic load: shall include seismic inertial force and earthquake hydrodynamic pressure on dam. The calculation may be performed in accordance with SL 203—97 *Specifications for Seismic Design of Hydraulic Structures*.

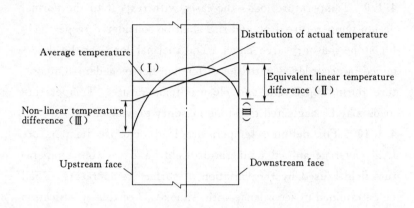

Figure 4.1.11 Distribution of dam temperature

4.2 Load Combinations

4.2.1 The load combinations in the design of stone masonry dams may be categorized as usual combination and unusual combination.

4.2.2 The load combinations for stone masonry gravity dams shall be determined according to the stipulations in SL 319—2005 *Design Standard for Concrete Gravity Dams*, as shown in Table 4.2.2 - 1. The load combinations for stone masonry arch dams are determined in accordance with Table 4.2.2 - 2.

Table 4.2.2-1 Load combinations for stone masonry gravity dam

Load combinations	Main load cases	Category of loads									Remarks
		Dead weight	Hydrostatic pressure	Uplift pressure	Silt pressure	Wave pressure	Ice pressure	Seismic loads	Hydrodynamic pressure	Earth pressure	Other loads	
Usual combination	Normal storage level	√	√	√	√	√				√	√	The earth pressure depends on whether there is earthfill against the dam
Usual combination	Design flood level	√	√	√	√	√			√	√	√	The earth pressure depends on whether there is earthfill against the dam
Usual combination	Freezing	√	√	√	√		√			√	√	The hydrostatic pressure and uplift pressures shall be calculated in accordance with the corresponding reservoir water level in winter
Unusual combination	Check flood level	√	√	√	√	√			√	√	√	
Unusual combination	Earthquake	√	√	√	√	√		√		√	√	The hydrostatic pressure, uplift pressure, and wave pressure shall be calculated in accordance with the normal storage level, which may be other water level if justified

Notes: 1. The most unfavorable loads combination shall be selected based on practicable possibility of simultaneous occurrence of various loads.
2. Dams to be constructed in stages shall have their corresponding load combination calculated for each stage.
3. Load combination in construction period shall be examined as an unusual load case.
4. If, according to geology or other conditions, the drainage system in the dam is susceptible to blocking and needs to be repaired during operation, load combination with drains inoperative shall be considered, as an unusual case.
5. For earthquake condition, if ice pressure in winter is considered, wave pressure shall be excluded.

Table 4.2.2 – 2 Load combinations for stone masonry arch dam

Loads combinations	Main load cases	Category of loads										
		Dead weight	Hydrostatic pressure	Uplift pressure	Silt pressure	Wave pressure	Ice pressure	Hydrodynamic pressure	Temperature loads		Seismic loads	Others
									Design normal temperature drop	Design normal temperature rise		
Usual combination	Normal storage level	✓	✓	✓	✓	✓	✓		✓			
	Design flood level	✓	✓	✓	✓	✓		✓		✓		
	Dead water level (or minimum operational water level) 1	✓	✓	✓	✓				✓			
	Dead water level (or minimum operational water level) 2	✓	✓	✓	✓	✓				✓		
	Other unfavorable cases of frequent loads	✓										✓

Table 4.2.2-2 (Continued)

Loads combinations	Main load cases		Dead weight	Hydrostatic pressure	Uplift pressure	Silt pressure	Wave pressure	Ice pressure	Hydrodynamic pressure	Temperature loads		Seismic loads	Others
										Design normal temperature drop	Design normal temperature rise		
Unusual combination	Check flood level		✓	✓	✓	✓	✓		✓		✓		
	Earthquake	Normal storage level + earthquake load	✓	✓	✓	✓	✓	✓		✓		✓	
		Dead water level 1 + earthquake load	✓	✓	✓	✓	✓	✓		✓		✓	
		Dead water level 2 + earthquake load	✓	✓	✓	✓	✓				✓	✓	
	Construction period		✓										
	Other unfavorable cases of infrequent loads		✓										✓

Notes: 1. In the above load combinations, controlling load combinations for computation may be figured out for calculation on the basis of actual engineering conditions.
2. Of wave pressure and ice pressure, the greater one shall be chosen in load combinations.
3. The load combination in construction period may be determined by taking into account the unfavorable conditions during actual construction.
4. Combinations of other frequent unfavorable loads and combinations of other infrequent unfavorable loads may be determined on the basis of study in accordance with specific engineering conditions.

5 Stone Masonry Gravity Dam

5.1 Dam Layout

5.1.1 The dam layout shall be carried out in combination with the project general layout in comprehensive considerations of the conditions such as topography, geology, and hydrology of dam site and the requirements for a multi-purpose project so as to locate rationally various component structures for flood release, power generation, irrigation, water supply, navigation, fish passage, floating debris removal, sediment flushing, reservoir emptying, downstream ecological water demand, etc. The optimal scheme shall be adopted through overall technical, and economic comparisons.

Attention shall also be paid to sediment flushing and bank slope protection in dam layout.

5.1.2 In the layout of dams, the dam type shall be chosen between concrete stone masonry gravity dam and cement mortar stone masonry gravity dam through technical and economic comparisons in accordance with the factors such as the distribution and quantity of local building materials, conditions for stone quarrying and shaping, and construction conditions.

5.1.3 The upstream length, bay numbers, types of weirs and piers, and elevation of weir crest of the overflow section shall be determined by taking into consideration the following factors:

   1 Requirements for flood regulation and floating debris removal.

   2 Topographical and geological conditions.

   3 Requirements for downstream water depth, energy dis-

sipation and scouring prevention.

4 Types and operation conditions of gates.

The optimal scheme shall be adopted through comprehensive comparisons of the above factors.

5.1.4 The facilities for flood releasing, energy dissipation and scouring protection shall be reasonably selected according to the topographical and geological conditions of downstream riverbed and banks, and the variation of downstream water depth, considering their influences on other structures.

5.1.5 For galleries, chambers and shafts to be arranged within the dam, their locations, types, elevations and sizes shall be reasonably determined to fulfill their functions and meet the requirements for operation and management.

5.1.6 The final scheme of dam layout for Class 2 structures shall be verified through hydraulic model tests, while that for Class 3 structures should be verified through hydraulic model tests.

5.2 Dam Structure

5.2.1 The dam structure shall be determined through extensive technical and economic comparisons in accordance with the loading conditions and the factors such as topography, geology, hydrology, building materials, and construction conditions.

5.2.2 The basic cross-section of non-overflow dam monoliths is essentially a triangle with its apex preferably above normal water level. The upper portion of the basic cross-section is shaped to accommodate the dam crest structure. The width of dam crest shall be determined according to the requirements of equipment layout, operation, transportation, etc.

5.2.3 For stone masonry gravity dams, the upstream faces of all dam monoliths should be consistent with each other, which is conducive to the connection and layout of impervious body. The downstream face of overflow section shall be separated by guide walls from that of non-overflow dam section.

5.2.4 The upstream face of solid gravity dams may be vertical, inclined or compounded. The upstream dam slope may range from 1 : 0.05 to 1 : 0.2. If a compounded face is used, the water intakes and the water release structures on the dam shall be arranged in coordination with each other. Downstream dam slope shall be determined according to the stress and stability requirements.

5.2.5 The structure layout of overflow dam face of stone masonry gravity dams may be determined according to the provisions in Section 4.3 of SL 319—2005 *Design Specification for Concrete Gravity Dams*. Overflow dam monoliths of stone masonry dams should be of ungated. If the overflow dam crest is gated, the structural stability and stresses of pier and gate chamber shall be analyzed, and the pier and gate chamber may be designed according to Section 6.5 of SL 319—2005.

5.2.6 The structure layout of outlet works through stone masonry gravity dam may be determined according to the provisions in Section 4.4 of SL 319—2005.

5.3 Calculation of Dam Sliding Stability

5.3.1 In the calculation of dam stability against sliding except deep sliding stability, the following cases shall be considered:

    1 Sliding along the interface between concrete pad and bedrock.

    2 Sliding along the interface between stone masonry and concrete pad.

    3 Sliding between masonry bodies.

5.3.2 The stability against of dam shall be calculated with shear-friction Formula (5.3.2-1) or friction Formula (5.3.2-2).

$$K' = \frac{f'\sum W + c'A}{\sum P} \quad (5.3.2-1)$$

$$K = \frac{f\sum W}{\sum P} \quad (5.3.2-2)$$

Where:

K' = safety factor of stability against sliding calculated by shear-friction formula;

f' = friction coefficient at the interface or section based on shear-friction formula;

c' = unit cohesion at the interface or section based on shear-friction formula, kPa;

A = area of slip plane, m^2;

$\sum W$ = summation of normal components of all forces on the slip plane, kN;

$\sum P$ = summation of tangential components of all forces on the slip plane, kN;

K = safety factor of stability against sliding calculated by sliding-friction formula;

f = friction coefficient of slip plane.

When Formula (5.3.2-1) or Formula (5.3.2-2) is used for calculation, the corresponding safety factor of sliding stability shall not be less than the values specified in Table 5.3.2.

Table 5.3.2 Safety factor of stability against sliding

Factor of safety	Formula	Load combinations		Class 2 or 3 dam
K'	(5.3.2-1)	Usual		3.00
		Unusual	1	2.50
			2	2.30

Table 5.3.2 (Continued)

Factor of safety	Formula	Load combinations		Class 2 or 3 dam
K	(5.3.2-2)	Usual		1.05
		Unusual	1	1.00
			2	1.00

5.3.3 When a weak structural plane or a gently dipping joint exist in the foundation rock, the sliding stability along these deep-seated planes shall be checked. According to the distribution of slip planes, the calculation modes may be classified into single-slip plane, double-slip plane, and multiple-slip plane modes. The calculations shall be done primarily with rigid body limit equilibrium method (refer to Annex E) or supplemented with the finite element method if necessary. The results from the comprehensive evaluation of calculations may form a basis for selecting the foundation treatment program. The formulas used for calculating stability against deep sliding are in Annex E of this specification.

5.3.4 For hollow gravity dams, in addition to calculating sliding stability of whole dam, the stability against sliding of the forelegs (the upstream solid part) shall also be checked.

5.4 Calculation of Dam Stresses

5.4.1 The methods for dam stress calculations shall meet the following requirements:

1 The gravity method shall be used as the basic method for the stress calculation of solid gravity dam and the corresponding formulas may be found in Annex C in SL 319—2005. When concrete slab is adopted on the dam face for seepage control, the layered elastic modulus method by considering unidirectional anisotropy of the dam may also be used for stress calculation as specified in Annex F of this specification.

  2 The finite element method should be adopted for stress calculation of high dams, dams on complex foundation and dams or dam monoliths which cannot be treated with plane models.

  3 The material mechanics method, structural mechanics method and finite element method may be used in the stress calculation of hollow gravity dam. The very unfavorable stress distribution shall not be allowed.

5.4.2 Stresses to be calculated mainly include:

  1 For a solid gravity dam, stresses on dam faces and the planes at which the dam slope changes shall be calculated. For medium and low dams, only the stresses of dam base may be calculated.

  2 For a hollow gravity dam, stresses around the spandrel arch and on upstream and downstream faces shall be calculated.

5.4.3 Safety factor for compressive strength of stone masonry gravity dams shall meet the following requirements:

  1 For usual load combinations, it shall not be less than 3.5.

  2 For unusual load combinations, it shall not be less than 3.0.

5.4.4 The dam stresses calculated by the material mechanics method shall meet the following requirements:

  1 For corresponding load combinations (except for seismic loads), the normal stress on dam base shall be less than the allowable compressive stress of stone masonry and the allowable bearing capacity of dam foundation. The allowable compressive stresses of stone masonry are specified in Table A.0.7 in Annex A of this specification.

2 The minimum normal stress on dam base shall be compressive stress.

3 The maximum principal compressive stress in dam shall be less than the allowable compressive stress of stone masonry.

4 Tensile stress shall not occur in dam except on overflow weir crest and the periphery of galleries and openings. Reinforced concrete structures shall be adopted where tensile stress occurs.

5 Gravity dam aseismic design, if necessary, shall conform to the provisions stipulated in SL 203—97.

5.4.5 For solid gravity dams, stresses of dam in construction period shall also be calculated, and the normal tensile stress at toe of the dam shall not exceed 100 kPa.

5.4.6 Stresses of hollow gravity dam shall meet the following requirements:

1 At dam heel: the principal tensile stresses are not allowed to occur at positions above 3%–5% of dam height from dam base (a smaller value should be chosen for high dams and a bigger value for medium and low dams).

2 At dam toe: the principal compressive stresses shall not be greater than the allowable compressive stress of stone masonry and the allowable bearing capacity of dam foundation.

5.4.7 For hollow gravity dams, measures shall be taken to adjust the shape of dam so as to reduce the tensile stress zones around the hollow space. The arch part should be of reinforced concrete structures.

5.4.8 The calculation of dam stresses by the finite element method may refer to the stipulations in Section 6.3 of SL 319—2005.

5.5 Dam Temperature Control and Crack Prevention

5.5.1 The design of temperature control shall be performed for high stone masonry gravity dams and medium dams of important projects, and temperature control measures shall be adopted. For large projects the finite element method should be used in the analyses of temperature fields and temperature stresses, and the calculation may refer to SL 319—2005. For medium and low dams of other projects, the temperature control and crack prevention may follow the experiences of similar projects.

5.5.2 The requirements for temperature control and crack prevention of foundation concrete may refer to SL 319—2005.

5.5.3 Joints shall be provided for stone masonry gravity dams. The arrangement of transverse joints shall be determined according to temperature stresses, dam layout, construction conditions, topographic conditions and geological conditions. The spacing between transverse joints should be 20 – 40 m. The position of transverse joints shall be adapted to foundation concrete construction joints. Longitudinal joints may not be provided for medium and low dams, while for high dams they should be provided according to temperature control.

5.5.4 The following measures should be taken to reduce temperature rise under the condition that the strength and durability of cementitious materials are not undermined:

    1 To use medium and low heat cements.
    2 To use fly ash or other admixtures.
    3 To use additives.

6 Stone Masonry Arch Dam

6.1 Dam Site, Dam Axis and Dam Layout

6.1.1 The site for stone masonry arch dams should be located on a comparatively narrow river valley, with relatively good geological conditions for abutments and thicker mountain mass at banks for the stability of arch abutments.

6.1.2 The location of the axis of an arch dam shall be selected by giving priority to the stability of arch abutments and through comparison of options.

6.1.3 The layout of arch dams shall be based on the natural conditions such as topography conditions, geology conditions, hydrology conditions, of dam site and the requirements for a multi-purpose project, and shall be determined through extensive comparison in terms of technology and economy.

6.1.4 The dam shall be reasonably shaped according to topography and geology of dam site, flood release modes, construction conditions, etc.

The maximum central angle of top arches should be 80° to 110°. For the dam site where the river valley is wide, non-circular arches should be selected. The overhang degree of cantilever for stone masonry arch dam should not be greater than 0.3 : 1.

6.1.5 The layout of flood release structures and the flood release mode for stone masonry arch dams shall be determined by technical and economic comparisons according to dam shape, dam height, flood discharge, the topography and geology of dam site, etc. For flood release structures integrated with the dam, priority should be given to surface spillway. Attention shall be paid to energy dissipation and scouring prevention for stone masonry arch dams. Hydraulic design shall conform to the provisions of *Design Specification for Concrete Arch Dams* SL 282—2003. The layout of flood release structures for Class 2 structures shall be verified by hydraulic model tests.

6.2 Dam Stress Analysis

6.2.1 In the dam stress analysis, the dam may be considered as an isotropic and homogeneous structure. If concrete impervious body is provided, the unidirectional anisotropy of dam may also be considered.

6.2.2 The trial load method should be used in the stress analysis of arch dam and the results should be used to evaluate the dam safety. For arch dams of Class 2 or stone masonry arch dams with complicated conditions, the trial load method shall be used for calculation and the finite element method shall be applied for verification if necessary.

6.2.3 Stress analysis mainly includes the followings:

    1 Stress distributions on the calculated cross-sections.

    2 Principal stresses at the calculation points on both upstream and downstream faces of dam.

    3 Local stresses in the weakened parts of dam (galleries, openings, etc.).

Some or all of the above-mentioned items shall be calculated based on specific conditions in different stages of design, and stresses inside dam foundation shall be analyzed if necessary.

6.2.4 In the dam stress analysis, the following issues shall be considered:

    1 Influence of dam shapes on stress distribution.

2 Influence of openings within dam on dam stresses.

3 Influence of closure temperature on dam stresses.

4 Influence of the dead weight of stone masonry arch dam on dam stresses when transverse joints are not provided.

5 Influence of staged construction and staged impoundment on dam stresses.

6 Influence of transverse joints on stresses and stability against overturning of individual dam monolith prior to grouting.

6.2.5 When the trial load method is used, the principal compressive stress and the principal tensile stress of dam shall meet the following requirement:

1 The safety factor for allowable compressive stress of stone masonry shall be 3.5 for basic load combinations, and 3.0 for special load combinations. The allowable compressive stress for stone masonry may be selected from Table A.0.7 when test data are unavailable.

2 For non-seismic load combinations, certain tensile stresses are allowable to occur in a stone masonry arch dam.

3 For seismic load combinations, the allowable tensile stress may be properly increased depending on specific conditions of the project.

The normal stresses of arch and beam shall conform to the stress indexes in this Clause when calculated with the crown cantilever method.

6.2.6 When calculated with the finite element method, the allowable stresses should be in accordance with the relevant provisions of SL 282—2003. The allowable tensile stresses under seismic load combinations other than non-seismic load combinations may be appropriately increased.

6.2.7 In the stress analysis of Class 2 stone masonry arch dam, the elastic modulus Poisson ratio of stone masonry, and the deformation modulus and elastic modulus of dam foundation shall be determined by tests. At the feasibility study stage, the empirical data of similar conditions may be adopted if the above data are not available.

6.2.8 For the stone masonry arch dams of important projects, the arch dam limit analysis method may be used for checking. When this method is used for checking, the safety factor of dam strength, K_J, is the ratio of the ultimate load to the design load. The value of K_J shall not be less than 3.2 for usual load combinations, and shall not be less than 2.9 for unusual load combinations.

6.3 Abutment Stability Analysis

6.3.1 The stability of abutments on two banks shall be analyzed and verified to meet the requirements by corresponding design stages of stone masonry arch dam.

6.3.2 When evaluating the stability of abutments, the slip plane shall be determined reasonably. The parameters of sliding-friction strength of the slip plane shall be obtained by tests for Class 2 stone masonry arch dams, while those for Class 3 stone masonry arch dams may be selected according to the experiential data from the projects with similar geological conditions when test data are unavailable.

6.3.3 In the sliding stability analysis of abutments, the limit equilibrium method is the main method and the finite element method may be used as a supplementary method if necessary.

The stability analysis of abutments shall be considered as a spatial problem to determine the safety factor of stability against sliding of the abutment as a whole. If the situation is simple and no complex slip plane exists, the safety factor of sliding stability may be calculated considering the abutment as series of planar layers.

6.3.4 When the sliding stability of abutments is analyzed with the limit equilibrium method, Formula (6.3.4 - 1) or Formula (6.3.4 - 2) shall be used:

$$K' = \frac{\sum(Nf' + c'A)}{\sum T} \qquad (6.3.4-1)$$

$$K = \frac{\sum(Nf)}{\sum T} \qquad (6.3.4-2)$$

Where:

K', K = safety factors of sliding calculated by shear-friction formula and slide-shear formula, respectively;

N = forces perpendicular to slip plane;

T = forces along slip plane;

A = area of calculated slip plane, m^2;

f' = friction coefficient for shear – friction on slip plane;

c = cohesion for shear – friction on slip plane, kPa;

f = friction coefficient for sliding-friction on slip plane.

The friction coefficient and cohesion of shear – friction, f' and c', shall be selected according to the corresponding peak strengths of the rock mass. The friction coefficient for sliding – friction, f, shall be the following characteristic values: limit of proportionality for brittle rocks; yield strength for plastic or brittle – plastic rocks; residual strength for rocks undergoing shear moving.

When Formula (6.3.4 - 1) is used for calculations, the corresponding safety factors shall not be less than the values specified in Table 6.3.4.

Table 6.3.4 Safety factor of stability against sliding

Factor of safety	Formula	Load combinations		Class of structure	
				2	3
K'	(6.3.4-1)	Usual		3.25	3.00
		Unusual	1	2.75	2.50
			2	2.25	2.00
K	(6.3.4-2)	Usual		1.40	1.30
		Unusual	1	1.20	1.10
			2	1.10	1.00

Notes: 1. Unusual load combination 1 includes 'Check flood level, Construction period, Combinations of other unfavorable rare loads' in Table 4.2.2-2.
2. Unusual load combination 2 refers to 'Earthquake' in Table 4.2.2-2.

6.3.5 If large faults or weak zones exist at the downstream of abutments, reinforcement measures shall be taken to control the amount of deformation, and the influence of abutment deformation on dam stresses shall be checked.

6.3.6 Effective measures shall be taken to reduce the seepage pressure on rocks so as to ensure the safety and stability of abutments.

6.3.7 For the gravity blocks and thrust blocks of stone masonry arch dams, the stability analysis shall meet the relevant provisions in this section and the calculations of stress and stability shall be in accordance with Annex G.

6.4 Temperature Control

6.4.1 In the construction of stone masonry arch dams, transverse gaps, wide or narrow, may be provided near arch abutments or at other appropriate locations if necessary. The width of wide gaps should be 0.8 - 1.2 m.

6.4.2 The closure temperature (average daily temperature during closure) for stone masonry arch dams shall be controlled below the average annual temperature but should not be less than 5℃. In severe cold areas the closure temperature of stone masonry arch dams shall be determined by special research.

For stone masonry arch dams without transverse joints, the average daily temperature during construction should be below the average annual temperature. And when it exceeds the average annual temperature, masonry construction should not be performed if no cooling measures are taken or the ambient temperature is lower than 5℃.

7 Dam Seepage Control

7.1 General Provisions

7.1.1 Dam seepage control facilities shall be determined in accordance with the factors such as local natural conditions, building materials, construction technology, and construction experience.

7.1.2 The following seepage control facilities may be adopted for a stone masonry dam:

    1 Seepage resistant concrete (reinforced concrete) face slab (layer) on the upstream dam face.

    2 Impervious concrete core in the stone masonry close to the upstream face.

    3 Seepage control by the dam itself.

    4 Other measures proved by practice or study.

7.1.3 Air entraining agent shall be added to concrete for frost-resisting. The types and amount of cement, admixture and additive as well as the water-cement ratio, mix proportion and gas content for suchconcrete shall be determined by tests.

7.2 Impervious Concrete Face Slab and Core

7.2.1 The seepage resistance grade of the concrete of impervious face slab and core shall be determined on the basis of the water head and in accordance with the provisions in SL/T 191—96 *Design Code for Hydraulic Concrete Structures*, which are shown in Table 7.2.1.

Table 7.2.1 Seepage resistance grade of concrete

Water head h (m)	<30	30-70	70-150
Seepage resistance grade	W4	W6	W8

7.2.2 For the concrete of impervious face slab and core with the outer protective layer whose thickness is less than twice the maximum freezing depth, the frost resistant grade shall be determined according to Table 7.2.2, on the basis of the factors such as climate zone, number of freeze-thaw cycles, ambient temperature of dam face, and freezing condition.

Table 7.2.2 Frost resistant grade of concrete

Climate zone	Severe cold		Cold		Mild
Number of annual freeze-thaw cycles	≥100	<100	≥100	<100	—
Part severely frozen and hard to repair	F300	F300	F300	F200	F100
Part severely frozen but possible to repair	F300	F200	F200	F150	F50
Part underwater perennially	F200	F150	F100	F100	F50

Notes: 1. The climate zones are divided by the coldest average monthly temperature (t_a): severe cold zone $t_a < -10°C$; cold zone $-10°C \leqslant t_a \leqslant -3°C$; mild zone $t_a > -3°C$.

2. The number of annual freeze-thaw cycles is the statistical value of the alternating times that the air temperature drops from above +3°C to below −3°C then rises back to above +3°C in a year, or the fluctuation times of the predetermined design water level in the period in which the average daily air temperature is below −3°C in a year, whichever is larger.

3. The frost resistant grade of concrete shall be determined by the quick freeze-thaw tests stipulated in SD 105—82, and may also be obtained with 90 d or 180 d age specimens.

7.2.3 The bottom thickness of impervious concrete face slab or core should be 1/30 − 1/60 of the maximum water head, and the top thickness shall not be less than 0.3 m.

7.2.4 Temperature reinforcement should be placed in the impervious concrete face slab and the joints should be well provided.

7.2.5 The connection of the impervious concrete face slab or core with the dam may be carried out through tie reinforcement or by making the surface of the adjacent masonry rough. The space between the impervious concrete core and the upstream dam face should be 0.5 – 2.0 m.

7.2.6 The impervious concrete face slab or core should be embedded 1 – 2 m deep into the foundation, and be connected to the seepage control facilities of dam foundation as an integral part.

7.3 Seepage Control of Dam Body

7.3.1 For a stone masonry dam that uses concrete with one – or two – graded aggregates as cementitious materials and vibrated mechanically, the dam body should be used for seepage resisting. In the meantime, water pressure test shall be performed in drill holes to check the permeability rate. If the permeability rate cannot meet the requirements, grouting shall be carried out.

7.3.2 For stone masonry dams below 50 m high, constructed of coarse stones bonded by cement mortar and the joints on the upstream face are pointed deeply with high strength cement mortar, the dam body should be used for seepage resisting.

7.3.3 When the dam body is used for seepage resisting, the design of seepage control for the connection between the dam and foundation shall be performed.

7.4 Transverse Joints, Water Stop, and Drainage

7.4.1 Expansion joints shall be provided in impervious concrete face slab of stone masonry gravity dam. The joint spacing should be 10 – 15 m. If transverse joints are provided in the dam body, joints in concrete face slab and core shall be coordinated with transverse joints arrangement.

7.4.2 The transverse joint spacing of impervious concrete face slab or core of masonry gravity dam should be 10 – 15 m, or should be determined by careful study and be consistent with the form and position of transverse joints on dam body.

7.4.3 Reliable water stops shall be installed in the expansion joints of impervious concrete face slab and core. Double – line water stops shall be installed in the impervious concrete face slab for high dams, and water stops for medium and low dams may be appropriately simplified.

Water stop materials may include annealed copper sheet, plastic strip, rubber strip, etc. They shall be selected reasonably in accordance with the factors including water head, climate conditions, construction conditions, etc.

7.4.4 Water stops of transverse joints should be embedded 30 – 50 cm into bedrock, using anchor bars to ensure the connection between concrete and bedrock if necessary.

7.4.5 Contact grouting, water stops, etc. may be used for water blocking at the contact surface between steep slope sections and bedrock.

7.4.6 Vertical drainage holes, if required, should be provided downstream the water stops in transverse joints for stone masonry gravity dams. They shall lead to the lower drainage galleries or the horizontal drainage system of the dam. The seepage water gathering into collecting wells may be drained into the downstream by gravity discharge or pump drainage.

8 Treatment of Dam Foundation

8.0.1 The design of foundation treatment of stone masonry dams shall be determined on the basis of comprehensive study on the factors such as geological conditions, the relationship between the foundation and its superstructure, general layout, construction method. The foundation treated shall satisfy the requirements of strength, stability, stiffness, seepage resisting and durability.

8.0.2 The dam base of stone masonry dams shall be determined by technical and economic comparison in consideration of dam stability, foundation stress, physical and mechanical properties of rocks, class of rocks, foundation deformation and stability, requirements of superstructures for the foundation, effect and workmanship of the foundation treatment, construction period, project cost, etc. In principle, after the foundation treatment, the excavation volume shall be reduced (as much as possible) while the requirements for foundation strength and stability are satisfied.

Dams higher than 100 m should be built on fresh, slightly weathered or the lower part of weak weathered bedrock. Dams between 50 m and 100 m in height should be built on slightly weathered to the middle part of weak weathered bedrock. Dams lower than 50 m in height should be built on the middle to upper part of weak weathered bedrock. These requirements should be appropriately less stringent for the dam monoliths resting on higher parts of the abutments.

8.0.3 For the karst region or the foundation with large fault fractured zones and weak interlayers, special design of foundation treatment shall be performed, and if necessary, the design for bedrock weathering control within a certain range of dam shall be performed.

Curtain grouting or cutoff walls shall be used in karst area for seepage resisting. The choice of which method to use shall be based on the conditions such as scale of karst caves, the permeability of fissures due to solution. When karst caves connecting the upstream and downstream exist in the dam foundation, if they are not buried deeply or construction conditions permit, excavation and concrete backfilling maybe conducted. When karst caves connecting the upstream and downstream or highly permeable solution fissures exist on the curtain axis, if they are buried deeply and unsuitable for open excavation, continuous cutoff walls may be formed for seepage control by connecting the adits which are excavated and backfilled with concrete one by one, or cutoff walls may also be formed for seepage control through trench type excavation and then concrete backfilling.

For large fault fractured zones and weak interlayers, it is preferable to adopt excavation, concrete replacement, concrete deep key-walls, concrete tuck, cutoff walls, cement grouting or superfine cement grouting, chemical grouting, etc.

8.0.4 The dam base of a stone masonry gravity dam monolith should not have great difference in elevation at its upstream and downstream sides and should preferably be slightly inclined toward upstream; and should be excavated into steps with a certain width in the direction parallel to the dam axis.

The rock faces on which the abutments are to be built should be excavated into radial planes; however, non-full-radial planes may be used in case of excessive excavation resulting from rather thick arch ends. The bank slope shall be parallel to the dam axis and shall not possess big slope angle and steps. The longitudinal slope of the whole rock faces used for dam foundation shall be smooth with no abrupt changes.

8.0.5 In the excavation design of dam foundation, requirements for blasting operations shall be prescribed. Concrete bedding should be set as required between the dam base and the stone masonry dam.

8.0.6 The design of foundation treatments of stone masonry gravity dams may refer to the relevant provisions of SL 319—2005. The design of foundation treatments of stone masonry arch dams may refer to the relevant provisions of SL 282—2003.

9 Dam Structure

9.1 Crest Layout and Transportation

9.1.1 The height difference between the elevation of the top of the wave wall at upstream of dam crest and the normal storage level or the check flood level of reservoir shall be calculated by Formula (9.1.1). Of the two results, the higher one shall be used to calculate the top elevation of the wave wall.

$$\Delta h = h_b + h_z + h_c \tag{9.1.1}$$

Where:

h_b = wave height, m, to be determined according to Annex C.4;

h_z = height difference between the center line of wave and the normal storage level or the check flood level of reservoir, m, to be determined according to Annex C.4;

h_c = free height, m, to be taken according to Table 9.1.1.

Table 9.1.1 Free height h_c

Water level	Safety class of dam	
	2	3
Normal storage level (m)	0.5	0.4
Check flood level (m)	0.4	0.3

9.1.2 The structure of dam crest shall meet the requirements of equipment arrangement, maintenance, transportation, and monitoring, with attention being paid to safety, applicability, economy, and aesthetics.

Access bridges or service bridges shall be provided above the crest of overflow sections if required.

The crest width should not be less than 3 m for non-overflow sections of dams over 50 m high.

9.1.3 The wave wall may be a stone masonry, concrete, or reinforced concrete structure. It shall be integrated with the dam with the two ends being connected, with the rock mass of abutment. The wall body shall have enough strength and may be 1.2 m in height.

9.1.4 Bridges may be set as required at the downstream face of the stone masonry dam.

9.2 Galleries and Openings

9.2.1 Galleries and openings shall be set as required inside the dam body, around which reinforced concrete structures may be adopted. Galleries and openings shall be arranged comprehensively, as far as possible at the locations with smaller stresses. If galleries and openings inside the dam body are of overhead crossing, the net spacing between them should not be less than 3 m. Galleries may not be provided inside a lower thin arch dam.

9.2.2 The distance between the upstream wall of longitudinal gallery and the upstream dam face should be 0.05 to 0.10 times of the water head acting on dam face, and shall not be shorter than 3 m.

The distance between the bottom surface of dam foundation grouting gallery and the bedrock surface shall not be less than 1.5 times of the width of gallery. The cross-sections of galleries should be in a shape of vault and straight sided walls. The galleries should be 2.5 - 3.0 m in width and 3.0 - 4.0 m in height. The bank slope of longitudinal gallery should not be steeper than 45°.

9.2.3 Foundation drainage galleries should be arranged on or near the bedrock surface. They should be 1.2 – 2.5 m in width and 2.2 – 3.0 m in height.

9.2.4 The layout of longitudinal inspection and observation galleries shall be compatible with the requirements of corresponding facilities. The inspection and observation galleries of hollow gravity dams should be connected with the hollow and led outside the dams.

9.2.5 When galleries at different elevations are provided within the dam, the elevation difference between two adjacent galleries should range from 20 – 40 m. The galleries at different elevations shall be connected with each other.

9.2.6 The galleries shall be provided with reliable lighting and drainage facilities, of which safety protection shall be ensured.

9.2.7 When there are monitoring instruments installed inside the dam, the observation rooms with cables leading into should be arranged outside the dam for the purpose of centralized observation.

9.3 Joints, Drainage, and Foundation Bedding

9.3.1 Settlement or temperature transverse joints may be provided according to the factors such as topography, geology, and temperature for stone masonry dams. Local construction joints may be provided if required. Water stops shall be installed in transverse joints of stone masonry gravity dam and horizontal joints of the foundation of stone masonry arch dam. The structure of transverse joints of stone masonry arch dam shall satisfy the requirements for arch closure grouting.

9.3.2 A row of vertical drain pipes should be provided in the dam. When seepage resistant facilities such as impervious slab and core are provided for the dam, the drain pipes shall be set downstream the facilities and the net spacing between them shall not be less than 2 m. When cutoff walls are not provided, the spacing between the drain pipes and the upstream dam face shall not be less than 3 m, and the drain pipes shall be compatible with the layout of drainage galleries. The drain pipes should be spaced about 3 - 5 m, their inner diameter should be about 15 cm, their upper ends shall extend to longitudinal galleries or the dam crest (with cover plate), and their lower ends shall extend to longitudinal inspection galleries or horizontal drain pipes. The elevation difference between the horizontal drain pipes should be 10 - 20 m.

Drain pipes may adopt precast no - sand concrete pipes or blind drains made of modified plastics. Drain pipes may be set as required on the contact surface of concrete overflow protection face and dam masonry and be extended towards the downstream.

9.3.3 Concrete cushion with a stiffness similar to that of adjacent stone masonry may be set as required on the dam base of stone masonry dam, and the thickness should be about 1 m.

10 Safety Monitoring Design

10.1 General Principles

10.1.1 A stone masonry dam shall be provided with necessary monitoring instruments or devices according to the class, height, structural type and features of dams, as well as the topographic and geological conditions of the site. The monitoring shall conform to the following stipulations:

1 Monitor, respectively, the status and safety of various structures during construction, reservoir impoundment and operation.

2 Verify the design and provide guidelines for construction.

3 Collect the data for scientific research.

10.1.2 The design of safety monitoring for stone masonry dams may refer to the stipulations of SDL 336—89 *Technical Specification for Concrete Dam Safety Monitoring (Trial Edition)*.

10.1.3 The scope of safety monitoring for stone masonry dams shall cover the dam, dam foundation and abutments, plus the bank slopes in the vicinity of the dam, which significantly affect the dam safety, as well as the structures and equipment that are directly associated with the dam safety.

10.1.4 The design of safety monitoring shall follow the principles listed below:

1 The design of safety monitoring shall be able to reflect completely and accurately the actual status of various structures in the periods of construction, reservoir impoundment and operation.

2 The monitoring items and the instrumentation layout shall be specifically selected and arranged in conjunction with the major factors that affect the safety of the project. For some important monitoring items in key sections or locations, it is preferable to adopt more than two methods to implement the monitoring.

3 The monitoring instruments and equipment to be adopted shall be reliable and stable with satisfying measuring range and accuracy. The monitoring methods used shall be of proven technique and convenient for operation.

4 Technologies adopted should be either advanced or be possible for future updating. Manual observation shall also be possible for locations with automatic monitoring system.

10.1.5 The design of safety monitoring shall meet the following requirements:

1 Special attention shall be paid to the safety monitoring in the periods of construction and initial reservoir impoundment, and the datum of major monitoring items shall be acquired timely. A detailed monitoring program shall be established prior to initial reservoir impoundment. Temporary monitoring facilities shall be correspondingly applied in the absence of monitoring conditions. During the impoundment, the measured data shall be sorted out and analyzed rapidly and fed back in time.

2 Besides the layout of safety monitoring facilities, the observation stations shall be arranged comprehensively as well, and the monitoring facilities shall be provided with good supporting conditions.

3 The monitoring facilities shall be embedded and installed in a way causing the least interference with construction activities. The instruments and cables should be reliably protected.

4 A predicted variation range of measured values for the major monitoring items should be defined on the basis of results obtained from theoretical calculations or model tests and according to the similar projects as well.

10.2 Monitoring Items and Monitoring Facility Layout

10.2.1 Necessary monitoring items shall be established for the safety monitoring of stone masonry dams. The monitoring shall be conducted by combining the means of instrumentation and patrol inspection.

10.2.2 The instrumentation monitoring is categorized as routine monitoring items and special monitoring items, both shall meet the following provisions.

1 The routine monitoring items include the monitoring of environmental variables, deformation, seepage, stress and strain, temperature, etc.

2 The special monitoring items refer to the items other than routine monitoring items and shall be defined by demonstration according to the need of projects.

10.2.3 The layout of monitoring facilities shall meet the following requirements:

1 The monitoring items of environmental variables mainly includes water level, reservoir water temperature, precipitation, sedimentation in front of the dam, downstream scouring, etc. The layout of monitoring facilities may refer to the stipulations of SDJ 336—89.

2 Horizontal displacement and deflection monitoring: The horizontal displacement of gravity dam and foundation should be monitored and measured with the tension wire alignment method. When the dam axis is short and the atmospheric conditions are favorable, the horizontal displacement of dam may be monitored and measured with the collimation method. The horizontal displacement of arch dam and foundation should be monitored and measured with the traverse method, or with the collimation method when having conditions. The surface horizontal displacement of dam slope may be monitored and measured with the triangulation network method, trilateration network method, intersection method and collimation method, etc. The horizontal displacement of the deep part of bank slopes in the vicinity of dam and that of the locations with geological defects such as faults and fissures may be monitored and measured with a group of inverted plumb lines, or electrical measuring instruments, such as bedrock deformation gauge, multipoint extensometer, borehole inclinometer. The deflection of dam should be measured by the plumb line method.

3 Vertical displacement monitoring: The vertical displacement of the dam and foundation should be measured with the precise leveling method, and the original datum marks for precise leveling should be set on the bedrock of bank slopes in the vicinity of dam. The datum benchmark for leveling shall be located at some place downstream of the dam that is free from the influence from the deformation of reservoir area.

4 Seepage monitoring: To measure the uplift pressure of dam foundation, piezometric tubes and osmometers may be installed along the longitudinal monitoring sections and transverse monitoring sections. The longitudinal monitoring section should be located along the first row of drainage curtain downstream of the anti-seepage curtain, with one measuring point laid in each dam monolith. The transverse monitoring sections should be located in the highest dam monolith, the abutment dam monolith and the monoliths with complicated geological conditions, and the measuring points shall not be fewer than 3 in each monitoring section. For thin arch dams with good geological conditions, monitoring facilities may not be provided for measuring the uplift pressure. In order to measure the seepage through the abutments, 2 to 3 monitoring sections may be provided downstream the grouting curtain along the seepage line in each abutment, and a minimum of 3 observation holes shall be drilled in each monitoring section. The observation holes shall be drilled deep into the highly pervious layer and well below the original groundwater table before the dam construction. In order to measure the seepage quantity through the dam body and foundation respectively, the measuring weirs should be installed at certain intervals in the foundation gallery drainage gutters.

5 Stress, strain and temperature monitoring: The sections used for monitoring the dam should be selected according to the height, length, shape, structure and geological conditions of the dam, and be arranged along the central section of monolith. For arch dams, the arch with the largest abutment stress shall be selected to locate a monitoring section, and monitoring instruments shall be centralized in every dam monolith. Temperature monitoring monoliths are key monoliths for the monitoring system. Thermometers shall not be provided where other instruments capable of monitoring temperature have been provided. Temperature measuring points should be arranged as a grid in the central sections of monitoring monoliths for gravity dams. Within the temperature monitoring monolith, 3 to 7 monitoring sections may be set according to the dam height. Temperature measuring points may be properly arranged at the downstream dam face subject to solar radiation.

6 The layout of monitoring facilities may refer to the provisions of SDL 336—89.

10.2.4 Patrol inspection shall meet the following requirements:

1 The dam and appurtenant structures shall be inspected through periodic patrol, from construction period to operation period. The frequency of the patrol inspection shall be increased during the period of initial reservoir impoundment or against the backdrop of rapid rise or sudden drawdown of reservoir water level, heavy flooding, felt earthquakes or other special events.

2 Any damage to the dam and its appurtenant structures or abnormality of the valley slopes in the vicinity of dam, the groundwater level, the foundation seepage and so on, shall be reported immediately and the observation and inspection shall be intensified. The causes shall be identified and remedial measures shall be studied for action.

3 Patrol inspection items may be defined according to the provisions of SDL 336—89.

Annex A Main Mechanical Indexes of Stone Masonry

A. 0. 1 Physical and mechanical test results of several stones are shown in Table A. 0. 1.

A. 0. 2 Values of the deformation modulus E_0 and elastic modulus E_e of stone masonry are shown in Table A. 0. 2.

A. 0. 3 Values of the ultimate axial compressive strength f_{cc} of stone masonry are shown in Table A. 0. 3.

A. 0. 4 Values of the ultimate tensile strength f_t of stone masonry are shown in Table A. 0. 4.

A. 0. 5 Reference values of shear – friction and sliding – friction strength parameters of the contact surface between concrete pad and bedrock for stone masonry dams are shown in Table A. 0. 5.

A. 0. 6 Reference values of shear – friction and sliding – friction strength parameters of the interface between stone masonry and concrete pad, or of the stone masonry itself are shown in Table A. 0. 6.

A. 0. 7 Values of the allowable compressive stress of stone masonry are shown in Table A. 0. 7.

Table A.0.1 Physical and mechanical test results of several stones

Name of stone	Dry density ρ_d (kg/m³)	Linear expansion coefficient α (10^{-6}/°C)	Ultimate strength (MPa)					Elastic modulus E_e (GPa)	Remarks
			Dry compressive R_d	Saturated compressive R_s	Tensile f_t	Bending f_t			
Sandstone	2100 – 2400	9.02 – 11.2	45 – 100	40 – 60	1.0 – 3.0	4 – 8		4 – 12	Mainly with reference to the test data of red sandstone in Sichuan
Limestone	2600 – 2800	6.75 – 6.77	110 – 150	80 – 140	4.0 – 6.0	13 – 28		50 – 70	Mainly with reference to the test data in Henan and Hunan
Granite	2500 – 2700	5.6 – 7.34	90 – 160	72 – 150	4.0 – 8.0	10 – 22		30 – 60	Mainly with reference to the test data in Hunan, Guangxi, and Shandong
Quartzite marble	2700 – 2800	6.5 – 10.12	100 – 120	80 – 100	4.5 – 6.0	6 – 16		20 – 30	Mainly with reference to the test data in Shaanxi

Table A.0.2 Values of the deformation modulus E_0 and elastic modulus E_e of stone masonry

Unit: GPa

Category of stone masonry	Saturated compressive strength R_s (MPa)	The specified strength of cementitious material (MPa)														
		Concrete						Cement mortar								
		≥15.0		10.0		12.5		10.0		7.5		5.0				
		E_0	E_e	E_0	E_e	E_0	E_e	E_0	E_e	E_0	E_e	E_0	E_e
Rubble	≥100	6.5	11.5	6.0	11.0	6.0	11.0	5.5	10.0	5.0	9.0	4.0	7.0
	80	6.0	11.0	5.0	9.0	5.0	9.0	4.5	8.0	4.0	7.0	3.0	5.5
	60	5.0	9.0	4.5	8.0	3.5	6.5	3.0	5.5	3.0	5.5	2.5	4.5
	50	4.0	7.0	4.0	7.0	2.0	3.5	2.0	3.5	2.0	3.5	2.0	3.5
	40	3.5	6.5	3.5	6.5	1.5	2.5	1.5	2.5	1.5	2.5	1.5	2.5
	30	3.0	5.5	3.0	5.5	1.0	2.0	1.0	2.0	1.0	2.0	1.0	2.0
70% rubble plus 30% ashlar	≥100	7.9	14.1	7.4	13.4	7.4	13.4	6.7	12.4	6.2	11.1	5.2	8.7
	80	6.9	12.5	6.1	11.0	6.1	11.0	5.6	10.0	4.9	8.7	3.9	7.2
	60	5.6	10.5	5.3	9.4	4.3	7.9	3.8	6.9	3.6	6.6	3.1	5.6
	50	4.3	7.6	4.3	7.6	2.5	4.4	2.3	4.1	2.3	4.1	2.3	4.1
	40	3.7	6.7	3.7	6.7	2.0	3.4	1.8	3.1	1.8	3.1	1.8	3.1
	30	3.2	5.8	3.2	5.8	1.3	2.5	1.3	2.5	1.3	2.5	1.3	2.5

Table A.0.2 (Continued)

Category of stone masonry	Saturated compressive strength R_s (MPa)	The specified strength of cementitious material (MPa)															
		Concrete						Cement mortar									
		≥15.0		10.0		12.5		10.0		7.5		5.0					
		E_0	E_e	E_0	E_e	E_0	E_e	E_0	E_e	E_0	E_e	E_0	E_e				
30% rubble plus 70% ashlar	≥100	9.7	17.5	9.2	16.6	9.2	16.6	8.7	15.6	7.8	13.9	6.1	10.9				
	80	8.1	14.5	7.5	13.6	7.5	13.6	7.0	12.6	6.1	10.9	5.1	9.4				
	60	6.4	11.5	6.3	9.4	4.3	9.7	4.8	8.7	4.4	8.0	3.9	7.0				
	50	4.7	8.4	4.7	8.4	3.1	5.6	2.7	4.9	2.7	4.9	2.7	4.9				
	40	3.9	6.9	3.9	6.9	2.6	4.6	2.2	3.9	2.2	3.9	2.2	3.9				
	30	3.4	6.2	3.4	6.2	1.7	3.1	1.7	3.1	1.7	3.1	1.7	3.1				
Ashlar	≥100	11.0	20.0	10.5	19.0	10.5	19.0	10.0	18.0	9.0	16.0	7.0	12.5				
	80	9.0	16.0	8.5	15.5	8.5	15.5	8.0	14.5	7.0	12.5	6.0	11.0				
	60	7.0	12.5	7.0	12.5	6.0	11.0	5.5	10.0	5.0	9.0	4.5	8.0				
	50	5.0	9.0	5.0	9.0	3.5	6.5	3.0	5.5	3.0	5.5	3.0	5.5				
	40	4.0	7.0	4.0	7.0	3.0	5.5	2.5	4.5	2.5	4.5	2.5	4.5				
	30	3.5	6.5	3.5	6.5	2.0	3.5	2.0	3.5	2.0	3.5	2.0	3.5				

Table A.0.2 (Continued)

Category of stone masonry	Saturated compressive strength R_s (MPa)	The specified strength of cementitious material (MPa)														
		Concrete						Cement mortar								
		≥15.0		10.0		12.5		10.0		7.5		5.0				
		E_0	E_e	E_0	E_e	E_0	E_e	E_0	E_e	E_0	E_e	E_0	E_e			
Coarse stone	≥100	10.0	18.0	9.5	17.0	9.5	17.0	9.0	16.0	8.0	14.5	7.0	12.5			
	80	8.0	14.5	8.0	14.5	7.5	13.5	7.0	12.5	6.5	11.5	5.5	10.0			
	60	7.5	13.5	7.0	12.5	6.5	11.5	6.0	11.0	5.5	10.0	4.5	8.0			
	50	6.5	11.5	6.0	11.0	5.5	10.0	5.0	9.0	4.5	8.0	4.0	7.0			
	40	5.5	10.0	5.0	9.0	4.0	7.0	4.0	7.0	4.0	7.0	3.5	6.5			
	30	4.0	7.0	4.0	7.0	3.0	5.5	3.0	5.5	3.0	5.5	3.0	5.5			

Note: Values of E_0 and E_e of the rubble masonry with concrete as the cementitious material, vibrated mechanically and laid with direct masonry method, may be increased by about 10% of the values for rubble masonry in Table A.0.2.

Table A.0.3 Values of the axial compressive strength f_{cc} of stone masonry

Unit: MPa

Category of stone masonry	Saturated compressive strength R_s (MPa)	The specified strength of cementitious material (MPa)						
			Concrete			Cement mortar		
		≥15.0	10.0	12.5	10.0	7.5	5.0
Rubble	≥100	14.4	11.2	11.2	9.6	8.0	6.8
	80	13.2	10.2	10.2	8.8	7.5	6.0
	60	11.6	8.8	8.8	7.6	6.4	5.2
	50	10.4	8.0	8.0	6.8	6.0	4.8
	40	9.2	7.0	7.0	6.0	5.2	4.4
	30	8.0	6.0	6.0	5.2	4.4	3.6
70% rubble plus 30% ashlar	≥100	17.3	13.5	13.5	11.5	9.6	8.1
	80	15.8	12.3	12.3	10.6	8.9	7.2
	60	13.9	10.6	10.6	9.2	7.7	6.3
	50	12.5	9.6	9.6	8.2	7.2	5.8
	40	10.8	8.4	8.4	7.2	6.3	5.2
	30	8.8	7.2	7.2	6.3	5.2	4.4

Table A.0.3 (Continued)

Category of stone masonry	Saturated compressive strength R_s (MPa)	The specified strength of cementitious material (MPa)						
		Concrete				Cement mortar		
		≥15.0	10.0	12.5	10.0	7.5	7.5	5.0
30% rubble plus 70% ashlar	≥100	21.2	16.5	16.5	14.1	11.6	9.9
	80	19.4	15.0	15.0	13.0	10.8	8.8
	60	16.9	13.0	13.0	11.2	9.5	7.7
	50	15.2	11.6	11.6	10.2	8.8	7.0
	40	13.0	10.4	10.4	8.8	7.7	6.4
	30	10.0	8.8	8.8	7.7	6.4	5.6
Ashlar	≥100	24.0	18.8	18.8	16.0	13.2	11.2
	80	22.0	17.1	17.1	14.8	12.2	10.0
	60	19.2	14.8	14.8	12.8	10.8	8.8
	50	17.3	13.2	13.2	11.6	10.0	8.0
	40	14.6	11.8	11.8	10.0	8.8	7.2
	30	10.8	10.0	10.0	8.8	7.0	6.4

Table A.0.3 (Continued)

Category of stone masonry	Saturated compressive strength R_s (MPa)	The specified strength of cementitious material (MPa)					Cement mortar	
		Concrete						
		≥15.0	10.0	12.5	10.0	7.5	5.0
Coarse stone	≥100	26.4	22.0	22.0	19.6	17.2	14.8
	80	24.4	19.9	19.9	18.0	15.6	13.7
	60	21.2	17.2	17.2	15.8	13.8	12.0
	50	18.9	15.3	15.3	14.4	12.8	10.8
	40	15.4	13.2	13.2	12.6	11.2	9.6
	30	10.8	10.8	10.8	10.8	9.6	8.4

Note: Values of f_{cc} of the rubble masonry with concrete as the cementitious material, vibrated mechanically and laid with direct masonry method may be increased to 105%–110%.

Table A.0.4 Values of the tensile strength f_t of stone masonry

Unit: MPa

Category	Failure mode	Category of stone masonry	Calculation method of f_t	The specified strength of cementitious material (MPa)					
				15.0	12.5	10.0	7.5	5.0	
Axial tensile	Along the continuous seam on the mortar joint interface of mortar joint	All	f_t	0.42	0.36	0.30	0.24	0.18	
	Along the slot seam on the interface of mortar joint	Rubble	$0.7 \times 2 f_t$	0.59	0.50	0.42	0.34	0.25	
		Coarse stone plus ashlar	$r \times 2 f_t$	0.84	0.72	0.60	0.48	0.36	
Bending tensile	Along the continuous seam on the interface of mortar joint	Various	$1.9 f_t$	0.80	0.68	0.57	0.46	0.34	
	Along the slot seam on the interface of mortar joint	Rubble	$1.9 \times 0.7 \times 2 f_t$	1.12	0.96	0.80	0.64	0.48	
		Coarse stone plus ashlar	$1.9 r \times 2 f_t$	1.60	1.37	1.14	0.91	0.68	

Notes: 1. r in the Table is the coefficient of bond, of which the value is equal to the ratio of stone bond length and masonry thickness of each layer. It is assumed in the Table that the bond length of rough dressed stone and ashlar masonry body is equal to masonry thickness of each layer. Therefore $r=1$. When $r \neq 1$, it shall be taken according to the actual situation.
2. Values of the ultimate tensile strength f_t when masonry body fails along the continuous seam on the interface of mortar joint are taken from ultimate tensile strength tests. Values of the ultimate tensile strength f_t of masonry body in other failure modes are calculated and obtained according to the calculation and obtaining value methods listed in the table.
3. Based on the results of pure bending tensile tests of rubble masonry body, values of f_t of the rubble masonry body vibrated mechanically and built with direct masonry method may be taken as about 110%.

Table A.0.5 Reference values of shear – friction and sliding – friction strength parameters of the interface between concrete pad and bedrock for stone masonry dams

Rock classification	Comprehensive evaluation of rock	Characteristics of dam foundation rock	Shear – friction strength parameters		Sliding – friction strength parameter
			f'	c' (MPa)	f
I	Very good	Complete, fresh, dense and hard, with cracks undeveloped, huge thick – bedded and thick – bedded rocks. Saturated compressive strength $R_s > 100$ MPa, deformation modulus $E_0 > 20$ GPa, P – wave velocity $V_p > 5$ km/s	1.2 – 1.5	1.3 – 1.5	0.7 – 0.75
II	Good	Complete, fresh, hard, with micro – fissures, thick – bedded rocks, saturated compressive strength R_s 60 – 100 MPa, deformation modulus E_0 10 – 20 GPa, P – wave velocity V_p 4 – 5 km/s	1.0 – 1.3	1.1 – 1.3	0.6 – 0.7
III	Medium	Low intactness, slightly weathered, medium hard, with micro – fissures, layered rock. Saturated compressive strength R_s 30 – 60 MP, deformation modulus E_0 5 – 10 GPa, P – wave velocity V_p 3 – 4 km/s	0.9 – 1.2	0.7 – 1.1	0.5 – 0.6

Table A. 0. 5 (Continued)

Rock classification	Comprehensive evaluation of rock	Characteristics of dam foundation rock	Shear-friction strength parameters		Sliding-friction strength parameter
			f'	c' (MPa)	f
IV	Poor	Poor integrity, weak weathered, relatively weak, with fissures, medium thick-bedded rocks, or with joints undeveloped, beddings and schistosity relatively developed, easy-weathered, thin-layered rock. Saturated compressive strength R_s 15-30 MPa, deformation modulus E_0 2-5 GPa, P-wave velocity V_p 2-3 km/s	0.7-0.9	0.3-0.7	0.35-0.5

Notes: 1. Bedrock with weak intercalated layers and weak structural planes is not included in the table. The rock in the table is bedrock of dam foundation.
2. The shear breaking strength parameters on the interface between bedding concrete and bedrock shall not be greater than those of bedding concrete itself, of which the specified strength is 15.0 MPa.
3. For rocks of Class I and II, if foundation planes fluctuate obviously, the shear-friction strength parameters on the interface may adopt the parameters of bedding concrete.

Table A.0.6 Reference values of shear – friction and sliding – friction strength parameters of the interface between stone masonry and concrete pad, or of stone masonry itself

Saturated compressive strength of stone masonry R_s (MPa)	Category of shear – friction and sliding – friction strength parameters	Cementitious materials (MPa)					
		Concrete		Cement mortar			
		15.0	10.0	12.5	10.0	7.5	5.0
>100	f'	1.1 – 1.4	1.0 – 1.3	1.0 – 1.3	0.9 – 1.2	0.8 – 1.0	0.7 – 0.9
	c' (MPa)	1.0 – 1.1	0.8 – 0.9	0.9 – 1.0	0.8 – 0.9	0.7 – 0.8	0.5 – 0.6
	f	0.65 – 0.75	0.65 – 0.75	0.65 – 0.75	0.65 – 0.75	0.55 – 0.65	0.5 – 0.6
60 – 100	f'	0.9 – 1.2	0.8 – 1.1	0.8 – 1.1	0.7 – 1.0	0.6 – 0.8	0.5 – 0.7
	c' (MPa)	0.8 – 1.0	0.6 – 0.7	0.7 – 0.8	0.6 – 0.7	0.5 – 0.6	0.4 – 0.5
	f	0.55 – 0.65	0.55 – 0.65	0.55 – 0.65	0.55 – 0.65	0.5 – 0.6	0.4 – 0.5
30 – 60	f'	0.8 – 1.1	0.7 – 0.9	0.7 – 0.9	0.6 – 0.8	0.5 – 0.7	0.4 – 0.6
	c' (MPa)	0.5 – 0.8	0.4 – 0.6	0.4 – 0.7	0.4 – 0.6	0.3 – 0.4	0.2 – 0.3
	f	0.45 – 0.55	0.45 – 0.55	0.45 – 0.55	0.45 – 0.55	0.4 – 0.5	0.3 – 0.4

Note: c' in the Table should be amended when used in accordance with the specific circumstances of projects.

Table A.0.7 Values of the allowable compressive stress of stone masonry   Unit: MPa

Category of masonry body	Saturated compressive strength R_s (MPa)	Usual load combination									Unusual load combination								
		The specified strength of cementitious material (MPa)									The specified strength of cementitious material (MPa)								
		Concrete			Cement mortar						Concrete			Cement mortar					
		15.0	10.0	12.5	10.0	7.5	5.0			15.0	10.0	12.5	10.0	7.5	5.0		
Rubble	≥100	5.1	4.0	4.0	3.4	2.9	2.4			6.0	4.7	4.7	4.0	3.3	2.8		
	80	4.7	3.6	3.6	3.1	2.6	2.1			5.5	4.2	4.2	3.7	3.0	2.5		
	60	4.1	3.1	3.1	2.7	2.3	1.9			4.8	3.7	3.7	3.2	2.7	2.2		
	50	3.7	2.9	2.9	2.4	2.1	1.7			4.3	3.3	3.3	2.8	2.5	2.0		
	40	3.3	2.4	2.4	2.1	1.9	1.6			3.8	2.8	2.8	2.5	2.2	1.8		
	30	2.9	2.1	2.1	1.9	1.6	1.3			3.3	2.5	2.5	2.2	1.8	1.5		
70% rubble plus 30% ashlar	≥100	6.2	4.8	4.8	4.1	3.4	2.9			7.2	5.6	5.6	4.8	4.0	3.4		
	80	5.7	4.3	4.3	3.8	3.1	2.6			6.6	5.1	5.1	4.5	3.6	3.1		
	60	4.9	3.7	3.7	3.3	2.8	2.3			5.8	4.5	4.5	3.8	3.2	2.7		
	50	4.4	3.4	3.4	2.9	2.5	2.1			5.1	4.0	4.0	3.4	3.0	2.4		
	40	3.8	2.9	2.9	2.6	2.3	1.9			4.5	3.5	3.5	3.0	2.7	2.2		
	30	3.2	2.6	2.6	2.3	1.9	1.6			3.7	3.0	3.0	2.7	2.2	1.9		

Table A.0.7 (Continued)

Category of masonry body	Saturated compressive strength R_s (MPa)	Usual load combination						Unusual load combination					
		The specified strength of cementitious material (MPa)						The specified strength of cementitious material (MPa)					
		Concrete		Cement mortar				Concrete		Cement mortar			
		15.0	10.0	12.5	10.0	7.5	5.0	15.0	10.0	12.5	10.0	7.5	5.0
30% rubble plus 70% ashlar	≥100	7.6	5.9	5.9	5.0	4.2	3.5	8.8	6.9	6.9	5.9	4.8	4.1
	80	7.0	5.3	5.3	4.6	3.8	3.2	8.1	6.2	6.2	5.5	4.4	3.8
	60	6.1	4.6	4.6	4.0	3.4	2.7	7.0	5.4	5.4	4.7	4.0	3.3
	50	5.4	4.2	4.2	3.6	3.2	2.5	6.3	4.8	4.8	4.2	3.7	2.9
	40	4.6	3.7	3.7	3.2	2.7	2.3	5.3	4.1	4.1	3.7	3.3	2.6
	30	3.6	3.2	3.2	2.7	2.3	2.0	4.1	3.7	3.7	3.3	2.6	2.3
Ashlar	≥100	8.6	6.7	6.7	5.7	4.7	4.0	10.0	7.8	7.8	6.7	5.5	4.7
	80	7.9	6.0	6.0	5.3	4.3	3.6	9.2	7.0	7.0	6.2	5.0	4.2
	60	6.9	5.2	5.2	4.6	3.9	3.1	8.0	6.2	6.2	5.3	4.5	3.7
	50	6.1	4.7	4.7	4.1	3.6	2.9	7.1	5.5	5.5	4.8	4.2	3.3
	40	5.1	4.2	4.2	3.6	3.1	2.6	6.0	4.9	4.9	4.2	3.7	3.0
	30	3.9	3.6	3.6	3.1	2.6	2.3	4.5	4.2	4.2	3.7	3.0	2.7

Table A. 0. 7 (Continued)

Category of masonry body	Saturated compressive strength R_s (MPa)	Usual load combination									Unusual load combination								
		The specified strength of cementitious material (MPa)									The specified strength of cementitious material (MPa)								
		Concrete			Cement mortar				Concrete			Cement mortar			
		15.0	10.0	12.5	10.0	7.5	5.0	15.0	10.0	12.5	10.0	7.5	5.0
Coarse stone	≥100	9.4	7.9	7.9	7.0	6.1	5.3	11.0	9.2	9.2	8.2	7.2	6.2
	80	8.7	7.0	7.0	6.4	5.6	4.9	10.2	8.2	8.2	7.5	6.5	5.5
	60	7.5	6.1	6.1	5.7	5.0	4.3	8.8	7.0	7.0	6.5	5.8	5.0
	50	6.6	5.5	5.5	5.1	4.6	3.9	7.8	6.2	6.2	6.0	5.3	4.5
	40	5.5	4.8	4.8	4.6	4.0	3.4	6.4	5.4	5.4	5.2	4.7	4.0
	30	3.9	3.9	3.9	3.9	3.4	3.0	4.5	4.5	4.5	4.5	4.0	3.5

Note: The values listed in the Table are obtained through the analysis of the test data of masonry body in 28 d - age design.

Annex B   Test Methods of Deformation (Elastic) Modulus and Compressive Strength of Stone Masonry

B.1   Test Purpose and Basic Principle

B.1.1   The purpose of deformation (elastic) modulus and compressive strength tests of stone masonry is to determine the compressive deformation (elastic) modulus and compressive strength of stone masonry.

B.1.2   The basic principle of deformation (elastic) modulus and compressive strength tests of stone masonry is that under the condition of keeping the cementitious materials and masonry method similar to the prototype, it is permissible to reduce the size of stones and make cube or prism specimens in a certain size for the tests. Under the condition that the shapes of the stones and those for dam construction are similar, the side length of the stones shall be less than 1/3 of that of specimens.

B.2   Specimen Size

B.2.1   The side length of specimens for laboratory tests shall not be less than 50 cm (the compression area in the tests shall not be less than 2500 cm^2). The side length of specimens for field tests in adits or test pits shall be greater than 70.7 cm (the compression area in the tests shall be greater than 5000 cm^2).

B.2.2   The ratio of specimen height to side length should be 1.2 – 1.5.

B.3   Specimen Masonry

B.3.1   In the same group of tests, there shall not be less than 5 specimens, and the cementitious materials and masonry method shall be consistent with construction. The specimens are prepared in a specially made test mold. When laid with the thick bed mortar method, the specimens should contain seven layers, including four layers of mortar and three layers of stones; or nine layers, including five layers of mortar and four layers of stones, if the ratio of specimen height to side length is 1.5. When using the direct masonry method, there shall be at least three layers of stones in a specimen.

B. 3. 2 In the meanwhile of preparing masonry specimens, a certain number of specimens of cementitious materials shall be prepared according to the requirements of standard specimen formation, and their compressive strengths shall be measured periodically prior to the tests.

B. 4　Test Equipment and Installation

B. 4. 1　The laboratory test is to place the ready made specimens on the 5000kN material testing machine. The test may be performed according to the methods specified in SD 105—82.

B. 4. 2　The equipment for field tests is mainly composed of loading system and measuring system. The loading system consists of hydraulic jack, electric oil pump, high - pressure oil pipes, pressure gauges, force transferring plates with sufficient rigidity, etc. The measuring system consists of displacement sensor (dial indicator), hydraulic pressure sensor, acoustic emission detector, nonmetal ultrasonic detector, function recorder, microcomputer, etc.

Loading equipment and measuring instruments are installed on the ready made specimens.

If the tests are conducted in test pits, specimens shall be upended and loading equipment and measuring instruments shall be installed horizontally.

B.5 Test Methods

B.5.1 The initial acoustic wave velocity of specimens shall be detected with ultrasonic detector before the test.

B.5.2 The test may adopt stepwise loading or one-step loading method. The pressure is increased first until the first appearance of acoustic emission signals or the moment when acoustic wave velocity starts to decrease and then dropped back to zero, and increased again at a constant rate up to the failure of specimen and the end of test. During the test, secondary instruments or microcomputers are used to take samples for monitoring the whole process of specimens under the action of forces. If the deformation is measured using a dial indicator, stepwise loading shall be adopted.

B.6 Test Results

B.6.1 Force value and deformation value shall be converted, respectively, to specific pressure σ and deformation amount s, and marked on $\sigma-s$ process curve drawn by recorder.

B.6.2 It is necessary to calculate acoustic wave velocity v_p and draw $\sigma-v_p$ curve graph.

B.6.3 Based on the correlation curves of $\sigma-s$, $\sigma-v_p$, $\sigma-A_E$ (acoustic emission rate), the feature points of three stages, i.e., initial cracking (proportional value), crack propagation (yield value) and rupture (peak value), may be determined accurately.

B.6.4 The deformation modulus E_0 and elastic modulus E_e of stone masonry shall be calculated respectively with Formula

(B. 6. 4 - 1) and Formula (B. 6. 4 - 2).

$$E_0 = \frac{P_2 - P_1}{A} \times \frac{L}{\Delta_{0L}} \times 10^{-5} \qquad \text{(B. 6. 4 - 1)}$$

$$E_e = \frac{P_2 - P_1}{A} \times \frac{L}{\Delta_{eL}} \times 10^{-5} \qquad \text{(B. 6. 4 - 2)}$$

Where:

E_0 = compressive deformation modulus, GPa;
E_e = compressive elastic modulus, GPa;
P_1 = compressive load, N;
P_2 = load when acoustic wave velocity begins to decrease or acoustic emission signals occur, N;
A = compression area of specimen, cm^2;
L = gauge length for measuring the deformation, cm;
Δ_{0L} = total deformation between P_1 and P_2, cm;
Δ_{eL} = elastic deformation between P_1 and P_2, cm.

B. 6. 5 Poisson ratio μ shall be calculated by Formula (B. 6. 5):

$$\mu = \frac{\varepsilon_t}{\varepsilon_l} \qquad \text{(B. 6. 5)}$$

Where:

μ = Poisson ratio of stone masonry;
ε_l = longitudinal strain of specimen;
ε_t = transverse strain of specimen.

B. 6. 6 Compressive strength f_{cc} of stone masonry shall be calculated by Formula (B. 6. 6):

$$f_{cc} = \frac{p_{max}}{A} \times 10^{-2} \qquad \text{(B. 6. 6)}$$

Where:

f_{cc} = compressive strength of stone masonry, MPa;
p_{max} = maximum load, N.

Annex C Calculation Formulas of Loads

C. 1 Hydrostatic Pressure

C. 1. 1 Intensity of hydrostatic pressure that acts perpendicularly at a certain point of dam face may be calculated by Formula (C. 1. 1):

$$p = \gamma_w H \qquad \text{(C. 1. 1)}$$

Where:

p = hydrostatic pressure intensity at a calculating point, kPa;

H = effective head at a calculating point, m;

γ_w = unit weight of water (or containing sediment), kN/m^3.

C. 1. 2 Horizontal hydrostatic pressure P that acts on the unit width of dam face may be calculated by Formula (C. 1. 2):

$$P = \frac{1}{2}\gamma_w H^2 \qquad \text{(C. 1. 2)}$$

Where:

P = horizontal hydrostatic pressure at a calculating point on the unit width, kN/m.

C. 2 Uplift Pressure

C. 2. 1 The calculation of uplift pressure that acts on the dam base shall conform to the following stipulations:

1 When impervious curtain and drainage holes are provided in dam foundation, the effective head of uplift pressure is H_1 (upstream water depth) at dam heel, $H_2 + \alpha(H_1 - H_2)$ at the central line of drainage holes and H_2 (downstream water depth) at dam toe, and those points are connected by a straight line. The reduction coefficient α may be determined according to geological conditions of dam foundation. Generally α can be 0.25 for sections on riverbed, 0.35 for sections on abutment [refer to Figure C.2.1a)]; the appropriate value of α shall be determined by study when the hydrogeology and engineering geology are complicated at banks.

2 When the geological conditions of dam foundation are good and only drainage holes but no impervious curtain are provided in dam foundation, the effective head of uplift pressure is H_1 at dam heel, $H_2 + \alpha_2(H_1 - H_2)$ at the central line of drainage holes and H_2 at dam toe, and those points are connected by a straight line. The reduction coefficient α_2 should be 0.3 to 0.45 [refer to Figure C.2.1b)].

3 When only impervious curtain but no drainage holes are provided in dam foundation, the effective head of uplift pressure is H_1 at dam heel, $H_2 + \alpha_1(H_1 - H_2)$ at the central line of curtain and H_2 at dam toe, and those points are connected by a straight line. The reduction coefficient α_1 could be 0.5 to 0.7 [refer to Figure C.2.1c)].

4 When neither impervious curtain nor drainage holes are provided in foundation, the effective head of uplift pressure is H_1 at dam heel and H_2 at dam toe, and those points are connected by a straight line [refer to Figure C.2.1d)].

C.2.2 Uplift pressure of hollow gravity dam foundation consists of two parts: one is the uplift pressure that acts between dam heel and the upstream side of the hollow, the value and figure of which are the same as shown in Figure C.2.1; the other is the uplift pressure that acts between the downstream side of the hollow and dam toe, which shall be determined respectively with Figure C.2.2a) or C.2.2b) based on whether there is water on the bedrock surface in the hollow.

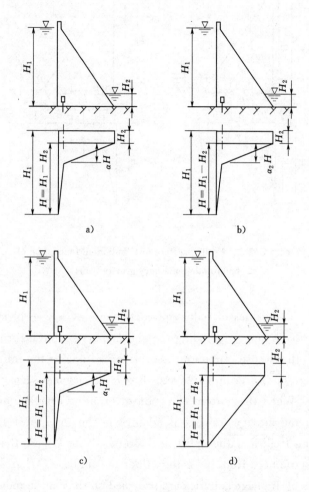

Figure C.2.1 Uplift pressure distribution on foundation of solid stone masonry dams

a) With impervious curtain and drainage holes; b) Without impervious curtain but drainage holes; c) With impervious curtain but no drainage holes; d) Without impervious curtain and drainage holes

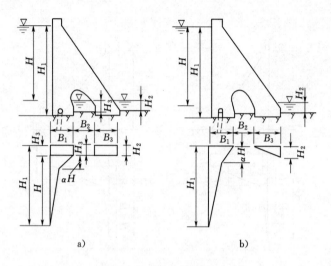

Figure C. 2. 2 Uplift pressure distribution on foundation of hollow stone masonry gravity dams

a) Water in the hollow; b) No water in the hollow

C. 2. 3 The values of reduction coefficients α, α_1, α_2 of monoliths of bank slope (including the abutments) should be properly greater than those of monoliths on riverbed due to the influence of the groundwater in banks and three-dimensional seepage.

C. 2. 4 When the bottom elevation of grouting drainage gallery is high and gravity drainage is adopted in the absence of special pumping facilities, the uplift pressure shall not be less than the hydrostatic head from the gallery floor to dam base. When several rows of drainage galleries are provided on the dam foundation, the effect of the first row of galleries may be considered only in design.

C. 2. 5 As for the calculation of the uplift pressure that acts inside solid gravity dams, the effective head of uplift pressure at the upstream dam face is the upstream water depth H'_1 which is above the calculating section, that at the central line of drainage pipe is $H'_2 + \alpha_3(H'_1 - H'_2)$ (when the calculating section is above the downstream water level, $H'_2 = 0$), that at the downstream dam face is the downstream water depth H'_2, which is above the calculating section, and those points are connected by a straight line. The reduction coefficient α_3 should be 0.15 to 0.25 [refer to Figure C.2.5a)]. When with no drainage pipes provided in dam, the effective head of uplift pressure at the upstream dam face is H'_1 and that at the downstream dam face is H'_2, and those points are connected with a straight line [refer to Figure C.2.5b)].

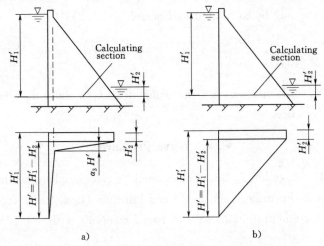

Figure C. 2. 5 Uplift pressure distribution in solid gravity dams

a) With drainage pipes; b) Without drainage pipes

C. 3 Silt Pressure

C. 3. 1 Horizontal silt pressure intensity p_{sk} acting on a unit width of dam face may be calculated according to the following formulas:

$$p_{sk} = \gamma_{sb} h_s \tan^2\left(45° - \frac{\varphi_s}{2}\right) \qquad (C.3.1-1)$$

$$\gamma_{sb} = \gamma_{sd} - (1-n)\gamma_w \qquad (C.3.1-2)$$

Where:

p_{sk} = silt pressure, kN/m;

γ_{sb} = submerged unit weight of sediment, kN/m³;

γ_{sd} = dry unit weight of sediment, kN/m³;

γ_w = unit weight of water, kN/m³;

n = porosity of sediment;

h_s = thickness of sediment deposited in front of the dam, m;

φ_s = internal friction angle of sediment, (°).

C.3.2 Horizontal silt pressure P_{sk} acting on a unit width of dam face may be calculated by Formula (C.3.2):

$$P_{sk} = \frac{1}{2}\gamma_{sb}h_s^2 \tan^2\left(45° - \frac{\varphi_s}{2}\right) \qquad (C.3.2)$$

Where:

P_{sk} = silt pressure, kN/m, acting at the $h_s/3$ above the foundation.

C.4 Wave Pressure

C.4.1 Wave height and length shall be calculated by Guanting Reservoir Formula (C.4.1-1) and Formula (C.4.1-2) (applicable for a mountainous valley-based reservoir with $v_0 < 20$ m/s and $D < 20$ km):

$$\frac{gh_b}{v_0^2} = 0.0076 v_0^{-\frac{1}{12}}\left(\frac{gD}{v_0^2}\right)^{\frac{1}{3}} \qquad (C.4.1-1)$$

$$\frac{gL_m}{v_0^2} = 0.33 v_0^{-\frac{7}{15}}\left(\frac{gD}{v_0^2}\right)^{\frac{4}{15}} \qquad (C.4.1-2)$$

Where:

h_b = wave height, m; when $\frac{gD}{v_0^2} = 20 - 250$, wave height at a cumulative frequency of 5% is $h_{5\%}$; when $\frac{gD}{v_0^2} = 250 - $

1000, wave height at a cumulative frequency of 10% is $h_{10\%}$;

L_m = average wave length, m;

v_0 = maximum wind speed for calculation, m/s; the maximum annual wind speed within a 50 - year return period can be used for the usual combinations and the long - term average of the maximum annual wind speed can be used for the unusual combinations;

D = fetch length, m;

g = acceleration of gravity, 9.81 m/s².

C.4.2 Wave pressure can be approximately calculated in accordance with a vertical wall type of retaining structure:

1 When water depth in front of dam is $H_1 \geqslant H_{cr}$, $H_1 > L_m/2$, the wave pressure distribution is given in Figure C.4.2a), and the wave pressure per unit width may be calculated according to the following formulas:

$$P_{wk} = \frac{1}{4}\gamma_w L_m (h_{5\%-10\%} + h_z) \qquad (C.4.2-1)$$

$$h_z = \frac{\pi h_{5\%-10\%}^2}{L_m} \coth \frac{2\pi H_1}{L_m} \qquad (C.4.2-2)$$

$$H_{cr} = \frac{L_m}{4\pi} \ln \frac{L_m + 2\pi h_{5\%-10\%}}{L_m - 2\pi h_{5\%-10\%}} \qquad (C.4.2-3)$$

Where:

P_{wk} = wave pressure per unit width on the dam face, kN/m;

$h_{5\%-10\%}$ = wave height at a cumulative frequency of 5% to 10%, m;

h_z = height difference between wave central line and calculating water level, m;

H_{cr} = critical water depth leading to wave breaking, m.

2 When water depth in front of dam is $H_1 \geqslant H_{cr}$, $H_1 \leqslant L_m/2$, the wave pressure distribution is given in Figure C.4.2

b), and the wave pressure per unit width may be calculated according to the following formulas:

$$p_{wk} = \frac{1}{2}[(h_{5\%-10\%} + h_z)(\gamma_w H_1 + p_{lf}) + H_1 p_{lf}]$$

(C. 4. 2 - 4)

$$p_{lf} = \frac{\gamma_w h_{5\%-10\%}}{\cosh \frac{2\pi H_1}{L_m}}$$

(C. 4. 2 - 5)

Where:

p_{lf} = residual strength of wave pressure at the dam bottom, kPa.

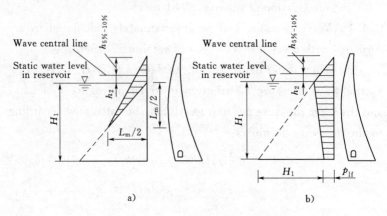

Figure C. 4. 2　Wave pressure distribution
a) $H_1 \geqslant H_{cr}$, $H_1 > L_m/2$; b) $H_1 \geqslant H_{cr}$, $H_1 \leqslant L_m/2$

C. 5　Ice Pressure

C. 5. 1　Static ice pressure acting on a unit-width of dam face may be retrieved from Table C. 5. 1.

C. 5. 2　The calculation of dynamic ice pressure shall conform to the following stipulations:

　　1　Dynamic ice pressure generated by ice striking on a vertical or nearly vertical dam face may be calculated by Formula

(C.5.2-1):

Table C.5.1 Static ice pressure values

Ice thickness (m)	0.4	0.6	0.8	1.0	1.2
Static ice pressure value (kN/m)	85	180	215	245	280

Notes: 1. Ice thickness is taken as the maximum annual average value.
   2. Static ice pressure values as shown in the table are available to the reservoir with a steady water level in a freezing period.
   3. For a small-sized reservoir, the static ice pressure value selected from the table shall be multiplied by 0.87.

$$F_{bk} = 0.07 v d_i \sqrt{A f_{ic}} \quad\quad (C.5.2-1)$$

Where:

F_{bk} = dynamic ice pressure generated by ice striking on dam face, MN;

v = flow speed of ice, m/s, to be determined by the data from observation; when the data is unavailable, it can be 3% of the maximum wind speed in an ice movement period, but not to be higher than 0.6 m/s;

d_i = calculating ice cover thickness, m, may be 0.7-0.8 times the maximum thickness of local ice, a bigger value is taken for the initial period of ice drifting;

A = area of ice, m², can be determined by the data from observation or investigation on site or adjacent areas;

f_{ic} = compressive strength of ice, MPa, is suitably determined by tests, when without the test data, it may be 0.3 MPa.

2 Dynamic ice pressure acting on a single pier or pile may be calculated by Formula (C.5.2-2):

$$F_{pl} = m f_{ib} d_i b \quad\quad (C.5.2-2)$$

Where:

F_{p1} = dynamic ice pressure generated by ice acting on a triangle pier or acting on a rectangular, polygonal or single circular pier with a vertical front, MN;

m = horizontal shape coefficient of the pier fronts, may be retrieved from Table C.5.2;

f_{ib} = crushing strength of the ice, MPa, may be 0.75 MPa for the initial period of ice drifting and be 0.45 MPa for the later period;

b = front width of the piers at an elevation where ice acting, m.

Table C.5.2 Shape coefficient m

Horizontal shape	Triangle with angle of 2α					Rectangular	Polygonal or circular
	45°	60°	75°	90°	120°		
m	0.54	0.59	0.64	0.69	0.77	1.0	0.9
Note: α is half of an angle of the triangle.							

C.6 Hydrodynamic Pressure

C.6.1 The hydrodynamic pressure on flip bucket section of overflow dam, assumed to be uniformly distributed, may be calculated by Formula (C.6.1):

$$p_0 = q\rho_w \frac{v}{R} \qquad (C.6.1)$$

Where:

p_0 = intensity of flow centrifugal pressure, Pa;

q = unit flow rate at inverse circle section under a corresponding designed flood release condition, m³/(s·m);

ρ_w = water density, kg/m³;

v = average flow speed in the section of the lowest point of flip bucket, m/s;

R = radius of flip bucket, m.

C. 6. 2 Horizontal and vertical components of resultant centrifugal force at a bucket of overflow dam may be calculated according to Formula (C. 6. 2 - 1) and Formula (C. 6. 2 - 2) respectively:

$$P_x = q\rho_w v(\cos\varphi_2 - \cos\varphi_1) \qquad (C.6.2-1)$$

$$P_y = q\rho_w v(\sin\varphi_2 + \sin\varphi_1) \qquad (C.6.2-2)$$

Where:

P_x = horizontal component of resultant centrifugal force on a unit width, N/m;

P_y = vertical component of resultant centrifugal force on a unit width, N/m;

φ_1, φ_2 = angle of bucket (refer to Figure C. 6. 2, absolute values taken).

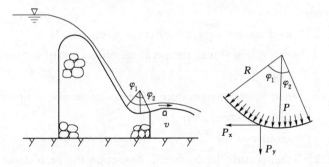

Figure C. 6. 2 Diagram of flow centrifugal force on flip bucket section

C. 6. 3 Centrifugal pressure intensity of flow that acts on the side wall of flip bucket section is assumed to be zero at the water surface and p_0 at the wall bottom; and it is required to approximately take linear distribution in the radial direction between those two locations. p_0 may be calculated by Formula (C. 6. 1) and acts perpendicularly on the wall surface.

C.7 Temperature Loads

C.7.1 The calculation of temperature loads for masonry arch dams shall conform to the following stipulations:

1 Temperature distribution in dam can be divided into three parts, i.e. average temperature T_m, equivalent linear temperature difference T_d and non-linear temperature difference T_n, as shown in Figure C.7.1. Those three may be calculated according to the following formulas respectively:

$$T_m = \frac{1}{L}\int_{-L/2}^{L/2} T\mathrm{d}x \qquad (C.7.1-1)$$

$$T_d = \frac{12}{L^2}\int_{-L/2}^{L/2} Tx\,\mathrm{d}x \qquad (C.7.1-2)$$

$$T_n = T - T_m - \frac{T_d x}{L} \qquad (C.7.1-3)$$

Where:

T_m = average temperature of cross-section, ℃;

T_d = equivalent linear temperature difference of cross-section, ℃;

T_n = non-linear temperature difference of cross-section, ℃;

L = dam thickness, m;

T = temperature, ℃, is a function of coordinate x.

2 Temperature loads during operation for arch dams may be calculated according to the following formulas respectively:

$$T_m = T_{m1} + T_{m2} - T_{m0} \qquad (C.7.1-4)$$

$$T_d = T_{d1} + T_{d2} - T_{d0} \qquad (C.7.1-5)$$

$$T_{m1} = \frac{1}{2}(T_{me} + T_{mi}) \qquad (C.7.1-6)$$

$$T_{d1} = T_{me} - T_{mi} \qquad (C.7.1-7)$$

Where:

T_{m0}, T_{d0} = average temperature and equivalent linear tem-

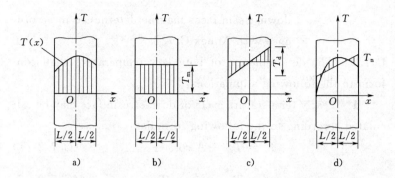

Figure C.7.1　Diagram of dam temperature distribution

a) Measured temperature; b) Average temperature T_m; c) Equivalent linear temperature difference T_d; d) Non-linear temperature difference T_n

perature difference of cross-section, which is determined by the closure temperature field of dam, may be calculated by Formulas (C.7.1-1) and (C.7.1-2) on the basis of actual temperature distribution in dam closing time;

T_{m1}, T_{d1} = average temperature and equivalent linear temperature difference of cross-section, which is determined by the average annual temperature field of dam, may be calculated by Formulas (C.7.1-6) and (C.7.1-7);

T_{m2}, T_{d2} = average temperature and equivalent linear temperature difference of cross-section, which is determined by the average annual temperature variation field of dam, may be determined in accordance with Annex C.7.3. When the closure temperature equals to the average annual temperature of dam, they are expressed by $T_{m0} = T_{m1}$, $T_{d0} = T_{d1}$, then $T_m = T_{m2}$, $T_d = T_{d2}$;

T_{mi}, T_{me} = average annual temperature of upstream and downstream faces may be determined in accordance with Annex C. 7. 2.

C. 7. 2 The determination of boundary temperature shall conform to the following stipulations:

1 Yearly temperature cycle of downstream face may be calculated according to the following formulas:

$$T_a = T_{am} + A_a \cos\omega(\tau - \tau_0) \quad \quad (C.7.2-1)$$

$$T_{am} = \frac{1}{12}\sum_{i=1}^{12} T_{ai} \quad \quad (C.7.2-2)$$

$$A_a = \frac{1}{2}(T_{a7} - T_{a1}) \quad \quad (C.7.2-3)$$

$$\omega = \frac{2\pi}{p_a} \quad \quad (C.7.2-4)$$

Where:

T_a = average monthly temperature, ℃;

T_{am} = average annual temperature, ℃, may be calculated by Formula (C. 7. 2 - 2) on the basis of measured average monthly temperature; in consideration of solar radiation, it is suitable to rise by 2 - 4℃;

A_a = annual amplitude of average annual temperature, ℃, may be calculated by Formula (C. 7. 2 - 3) on the basis of average monthly temperature observed in local area; in consideration of solar radiation it is suitable to rise by 1 - 2℃, and that below tail water level equals to annual amplitude of water temperature;

τ = time variable, month;

τ_0 = initial phase of time periodic variation process, month, it may be τ_0 = 6. 5 month;

ω = circular frequency, may be calculated by Formula (C. 7. 2 - 4);

p_a = temperature variation cycle, month, $p_a = 12$ month;

T_{ai} = average monthly temperature of i month, ℃;

T_{a1}, T_{a7} = average monthly temperatures of January and July, ℃.

2 Yearly temperature cycle of upstream face may be determined according to the following stipulations:

  1) Temperature above water surface equals to air temperature and may be calculated by Formula (C.7.2-1).

  2) Yearly variation of water temperature in front of dam may be expressed by Formula (C.7.2-5) in the preliminary design, and may be determined by studying the measured water temperature data of similar reservoirs in the final design.

$$T_w(y,\tau) = T_{wm}(y) + A_w(y)\cos\omega[\tau - \tau_0 - \varepsilon(y)]$$

(C.7.2-5)

Where:

  $T_w(y, \tau)$ = average annual water temperature at a water depth of y and in a time of τ, ℃;

  y = water depth, m;

  $T_{wm}(y)$ = average annual water temperature at a water depth of y is determined by Item 3 of this clause, ℃;

  $A_w(y)$ = annual amplitude of average annual water temperature at a water depth of y is determined by Item 4 of this clause;

  $\varepsilon(y)$ = phase difference between yearly periodic variation of water temperature at a water depth of y and that of air temperature, is determined by Item 5 of this clause, month.

3 Average annual water temperature at a water depth of y may be calculated according to the following formulas:

$$T_{wm}(y) = c + (b-c)e^{-0.04y} \quad \text{(C.7.2-6)}$$

$$b = T_{am} + \Delta b \quad \text{(C.7.2-7)}$$

$$c = \frac{T_{kd} - bg}{1 - g} \quad \text{(C.7.2-8)}$$

$$g = e^{-0.04H} \quad \text{(C.7.2-9)}$$

Where:

$b=$ average annual water temperature, ℃, may be calculated by Formula (C.7.2-2) (but in cold areas, when multi-year average monthly temperature $T_{ai} < 0$, then $T_{ai} = 0$);

$\Delta b=$ temperature increment caused by solar radiation, ℃, generally $\Delta b = 2℃$;

$c=$ first term value of average annual water temperature at a water depth of y, ℃;

$T_{kd}=$ average annual water temperature at the reservoir bottom, ℃ (in general areas is approximately equal to the minimum average three-month temperature, in severe cold areas is approximately equal to 4-6℃), may be retrieved from Table C.7.2 in the preliminary design;

$g=$ intermediate variable which the calculation requires, $g = e^{-0.04H}$;

$H=$ depth of reservoir, m.

Table C.7.2 Values of T_{kd}

Climate condition	Severe cold (Northeast)	Cold (North, Northwest)	Mild (Central, East, Southwest)	Torridity (South)
T_{kd} (℃)	4-6	6-7	7-10	0-12

In the calculation by Formula (C.7.2-6), if $T_{wm}(y) < T_{kd}$, then take $T_{wm}(y) = T_{kd}$.

4 The annual amplitude of average annual water temperature at a water depth of y, $A_w(y)$, may be calculated by Formula

(C.7.2-10):

$$A_w(y) = A'_a e^{-0.018y} \quad (C.7.2-10)$$

Where:

A'_a = modified annual amplitude of temperature, i. e. annual amplitude of surface water temperature, ℃ (when T_{am} <10℃, $A'_a = T_{a7}/2 + \Delta a$, when $T_{am} \geq 10℃$, $A'_a = A_a$);

Δa = temperature increment caused by solar radiation, ℃, generally $\Delta a = 1 - 2℃$.

5 Phase difference $\varepsilon(y)$ between yearly periodic variation of water temperature and that of air temperature at a water depth of y may be calculated by Formula (C.7.2-11):

$$\varepsilon(y) = 2.15 - 1.30 e^{-0.085y} \quad (C.7.2-11)$$

C.7.3 The calculation of average temperature T_{m2} and equivalent linear temperature difference of cross-section T_{d2} of cross-section in the temperature field of average yearly variation shall conform to the following stipulations:

Average temperature and equivalent linear temperature difference of cross-section, which is determined by the average annual temperature variation field of dam, may be determined in accordance with Annex C.7.3.

1 The calculation of T_{m2} and T_{d2} in temperature field of average yearly variation may be calculated according to the following formulas:

$$T_{m2} = \frac{\rho_1}{2}[A_e\cos\omega(\tau-\theta_1-\tau_0) + A_i\cos\omega(\tau-\theta_1-\varepsilon-\tau_0)]$$

$$(C.7.3-1)$$

$$T_{d2} = \rho_2[A_e\cos\omega(\tau-\theta_2-\tau_0) - A_i\cos\omega(\tau-\theta_2-\varepsilon-\tau_0)]$$

$$(C.7.3-2)$$

$$\rho_1 = \frac{1}{\eta}\sqrt{\frac{2(\cosh\eta - \cos\eta)}{\cosh\eta + \cos\eta}} \quad (C.7.3-3)$$

$$\rho_2 = \sqrt{a_1^2 + b_1^2} \qquad (C.7.3-4)$$

$$\theta_1 = \frac{1}{\omega}\left(\frac{\pi}{4} - \arctan\frac{\sin\eta}{\sinh\eta}\right) \qquad (C.7.3-5)$$

$$\theta_2 = \frac{1}{\omega}\arctan\frac{b_1}{a_1} \qquad (C.7.3-6)$$

$$a_1 = \frac{6}{\rho_1 \eta^2}\sin(\omega\theta_1) \qquad (C.7.3-7)$$

$$b_1 = \frac{6}{\rho_1 \eta^2}[\cos(\omega\theta_1) - \rho_1] \qquad (C.7.3-8)$$

$$\eta = \sqrt{\frac{\pi}{a_c p}L} \qquad (C.7.3-9)$$

Where:

ρ_1 = coefficient used to calculate T_{m2}, a function of η;

ρ_2 = coefficient used to calculate T_{d2}, a function of a_1 and b_1 which are functions of η, ρ_1, $\omega\theta_1$;

θ_1 = maximum time of T_{m2}, month;

θ_2 = a large time of T_{d2} corresponding to the maximum time of T_{m2} θ_1, month;

a_1 = one of the variables to calculate θ_2 corresponding to the maximum θ_1;

b_1 = the other variable to calculate θ_2 corresponding to the maximum θ_1;

a_c = temperature diffusivity of stone masonry, m²/month, related to masonry materials (for masonry of general granite and quartz sandstone, $a_c = 3.52$; for masonry of limestone, $a_c = 4.32$; for masonry of basalt and rhyolite, $a_c = 2.52$);

ε = phase difference between yearly periodic variation of temperature on upstream and downstream dam faces, in month, when water is on upstream face and air on downstream face, may be determined in accordance with

Annex C. 7. 2;

A_i, A_e = annual amplitude of average annual temperature at upstream and downstream dam faces, ℃;

τ = a calculating time point, with the worst combination of temperature loads, can be $\tau=7.5$ (mid August) or 8.0 (the end of August) and T_{m2} and T_{d2} which are corresponding to temperature increment, and it is allowable to change the sign to be T_{m2} and T_{d2} which are corresponding to temperature drop.

2 T_{m2} and T_{d2} of temperature field of average yearly variation may be simplified for calculation in the preliminary design:

1) Above water level of reservoir:

$$T_{m2} = \pm \rho_1 A_e \qquad (C.7.3-10)$$

$$T_{d2} = 0 \qquad (C.7.3-11)$$

2) Below water level of reservoir:

$$T_{m2} = \pm \frac{\rho_1}{2}\left(A_e + \frac{13.1 A_a}{14.5 + y}\right) \qquad (C.7.3-12)$$

$$T_{d2} = \pm \rho_3 \left[A_e - A_a\left(\xi + \frac{13.1}{14.5 + y}\right)\right]$$

$$(C.7.3-13)$$

3) when dam thickness $L \geqslant 10$ m:

$$\frac{\rho_1}{2} = \frac{2.33}{L - 0.90} \qquad (C.7.3-14)$$

$$\rho_3 = \frac{18.76}{L + 12.6} \qquad (C.7.3-15)$$

$$\xi = \frac{3.80 e^{-0.022y} - 2.38 e^{-0.081y}}{L - 4.50} \qquad (C.7.3-16)$$

4) when $L < 10$ m:

$$\frac{\rho_1}{2} = 0.50 e^{-0.00067 L^{3.0}} \qquad (C.7.3-17)$$

$$\rho_3 = e^{-0.00186 L^{2.0}} \qquad (C.7.3-18)$$

$$\xi = (0.069e^{-0.022y} - 0.0432e^{-0.081y})L \quad (C.7.3-19)$$

Where:

ρ_3 = simplified value that is equivalent to ρ_2;

ξ = first term related to A_a in the simplified expression for T_{d2}.

In Formulas (C. 7. 3 - 10) to (C. 7. 3 - 13), values are positive for summer and negative for winter. The formulas are used to calculate the temperature loads at the end of August and the end of February.

Annex D   Time Limitation for Calculations of Sedimentation in Front of Dam

The design reference period of main water retaining structures in hydraulic and hydroelectric engineering is 50 to 100 years. According to the design codes for gravity dams, arch dams, etc., when determining dam loads, the time limitation for calculation of sedimentation elevation in front of dams is 50 to 100 years.

The time limitation for calculations of sedimentation in reservoirs shall be determined in accordance with the following stipulations:

1   When the time limitation for the relative balance of sediment deposition in reservoirs is longer than the design reference period of water retaining structures, calculate to the reference period.

2   When the time limitation for the relative balance of sediment deposition in reservoirs is shorter than the design reference period of water retaining structures, calculate to the time limitation for the relative balance.

The time limitation for the relative balance of sediment deposition in reservoirs may be analyzed and judged by the conditions such as the average annual storage output rate of suspended load exceeding 90%, or the amount of sediment accumulated in reservoirs becoming almost stable.

Annex E Safety Factors of Deep Sliding Stability and the Formulas for Stone Masonry Material Mechanics Dam Foundation

E. 0. 1 When sub-horizontal potential failure planes exist in deep zone of a dam foundation, stability analysis against sliding shall be performed. The potential failure planes may be grouped as single slip plane, double slip plane and multiple slip plane according to geological conditions.

The situation of double slip plane is frequently encountered, as shown in Figure E. 0. 1.

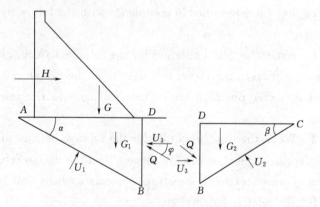

Figure E. 0. 1 Sketch of double slip plane

The safety factor of deep stability against sliding shall be calculated either by shear-friction strength formulas, i.e. Formulas (E. 0. 2 - 1) and (E. 0. 2 - 2), or by sliding-friction strength formulas, i.e. Formulas (E. 0. 3 - 1) and (E. 0. 3 - 2), based on the identical factor of safety method.

E. 0. 2 In the calculation with shear-friction strength formulas, considering the stability of wedge ABD, there is:

$$K_1' = \frac{f_1'[(W+G_1)\cos\alpha - H\sin\alpha - Q\sin(\varphi-\alpha) - U_1 + U_3\sin\alpha] + c_1'A_1}{(W+G_1)\sin\alpha + H\cos\alpha - U_3\cos\alpha - Q\cos(\varphi-\alpha)}$$

(E. 0. 2 - 1)

While considering the stability of wedge BCD, there is:

$$K_2' = \frac{f_2'[G_2\cos\beta + Q\sin(\varphi+\beta) - U_2 + U_3\sin\beta] + c_2'A_2}{Q\cos(\varphi+\beta) - G_2\sin\beta + U_3\cos\beta}$$

(E. 0. 2 - 2)

Where:

K_1', K_2' = safety factors of sliding stability calculated by shear-friction formulas;

W = vertical component of all loads acting on dam, kN;

H = horizontal component of all loads acting on dam, kN;

G_1, G_2 = vertical forces by weight of rock wedges ABD and BCD, respectively, kN;

f_1', f_2' = friction coefficients for shear-friction on slip planes AB and BC, respectively;

c_1', c_2' = cohesions for shear-friction on slip planes AB and BC, respectively, kPa;

A_1, A_2 = areas of slip planes AB and BC, respectively, m^2;

α, β = angles between slip planes AB and BC and horizontal, respectively, (°); if there is no obvious second slip plane in the downstream rock mass, the second slip plane may be determined by trial calculations, i.e. among a group of assumed β angles, the one corresponding to a slip plane with the smallest safety factor is the β angle of the second slip plane;

U_1, U_2, U_3 = uplift pressures acting on slip planes AB, BC and BD, respectively, kN;

Q, φ = force on slip plane BD, kN, and angle between the direction of force Q and horizontal which needs to be selected after verification and may be 0° conservatively.

Through Formula (E. 0. 2 - 1), Formula (E. 0. 2 - 2), and $K'_1 = K'_2 = K'$, the values of Q and K' can be solved.

E. 0. 3 For dam blocks where the safety factors derived by shear - friction strength formulas cannot meet the requirements as specified in Table 5. 3. 2 even after taking into account the proposed foundation treatment measures, the factors of safety against sliding may be calculated by sliding - friction strength formulas, i. e. Formulas (E. 0. 3 - 1) and (E. 0. 3 - 2), which shall not be less than the values specified in Table E. 0. 3.

Table E. 0. 3 Safety factor of stability against sliding K

Factor of safety	Load combination		Class of structures	
			2	3
K	Usual		1.30	1.15
	Unusual	1	1.15	1.10
		2	1.05	1.05

Considering the stability of wedge ABD, there is:

$$K_1 = \frac{f_1[(W+G_1)\cos\alpha - H\sin\alpha - Q\sin(\varphi-\alpha) - U_1 + U_3\sin\alpha]}{(W+G_1)\sin\alpha + H\cos\alpha - U_3\cos\alpha - Q\cos(\varphi-\alpha)}$$

(E. 0. 3 - 1)

Considering the stability of wedge BCD, there is:

$$K_2 = \frac{f_2[G_2\cos\beta + Q\sin(\varphi+\beta) - U_2 + U_3\sin\beta]}{Q\cos(\varphi+\beta) - G_2\sin\beta + U_3\cos\beta}$$

(E. 0. 3 - 2)

Where:

K_1, K_2 = safety factors of stability against sliding for sliding

— friction strength;

f_1, f_2 = friction coefficients of sliding – friction strength on slip planes AB and BC, respectively.

Through Formula (E. 0. 3 – 1), Formula (E. 0. 3 – 2), and $K_1 = K_2 = K$, the values of Q and K can be solved.

E. 0. 4 The situation of multi – slip plane is comparatively complicated. The value of K or K' for sliding along multi – slip planes may be solved by deriving the equilibrium formula for each wedge using for reference the formulas for the double slip plane.

Annex F Stress Calculations of Stone Masonry Gravity Dam with the Gravity Method, Considering the Distinct Elastic Moduli of Face Slab and Stone Masonry

F.1 Calculation Instructions

For the stone masonry gravity dams that the elastic modulus of impervious concrete slab differs significantly with that of stone masonry, the dam stresses may be calculated by the following method:

On the basis of linear change of the original vertical normal stress σ_y, the impervious concrete slab is first expanded by E_c/E_s times, with a method similar to the area transformation of rebar in reinforced concrete, to reflect the modulus difference, where E_c is the elastic modulus of impervious concrete slab and E_s that of stone masonry dam. Then according to the deformation compatibility on the interface between impervious slab and dam body and on the assumption that both Poisson's ratios are equal, the vertical normal stress σ_y of the expanded impervious slab can be obtained by the elasticity theory. It times the stress amplification coefficient E_c/E_s is the actual stress of impervious slab of stone masonry gravity dams.

Considering the characteristic of distinct elastic moduli of dam layers, the dam should be converted into a homogeneous body by the equivalent section approach in the calculations.

F.2 Calculation of Related Parameters

The section with layers of distinct elastic moduli is shown in

Figure F. 2.

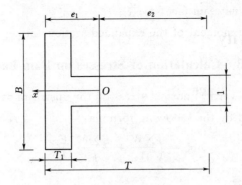

Figure F. 2

The related parameters may be calculated by the following formulas:

$$B = \frac{E_c}{E_s} \quad (F.2-1)$$

$$A = T + (B-1)T_1 \quad (F.2-2)$$

$$e_1 = \frac{1}{2} \frac{T^2 + (B-1)T_1^2}{A} \quad (F.2-3)$$

$$e_2 = T - e_1 \quad (F.2-4)$$

$$J = \frac{1}{3}[Be_1^3 - (B-1)(e_1 - T_1)^3 + e_1^3] \quad (F.2-5)$$

Where:

B = stress amplification coefficient;

E_c = elastics modulus of the impervious concrete slab, GPa;

E_s = elastics modulus of the stone masonry dam, GPa;

A = equivalent area, m^2;

T = thickness of the dam, m;

T_1 = thickness of the impervious slab, m;

e_1 = distance from the centroid of equivalent area to the upstream face, m;

e_2 = distance from the centroid of equivalent area and the downstream face, m;

J = area moment of the expanded section, m³.

F.3 Calculation of Stresses on Dam Face

F.3.1 The vertical normal stress on the upstream dam face may be calculated by the following formula:

$$\sigma_y^s = \left(\frac{\sum W}{A} + \frac{e_1 \sum M}{J}\right)\frac{E_c}{E_s} \qquad (F.3.1)$$

Where:

σ_y^s = vertical normal stress on the upstream dam face, MPa.

F.3.2 The vertical normal stress on the downstream dam face may be calculated by the following formula:

$$\sigma_y^{xi} = \frac{\sum W}{A} - \frac{e_2 \sum M}{J} \qquad (F.3.2)$$

Where:

σ_y^{xi} = vertical normal stress on the downstream dam face, MPa.

F.3.3 The shear stress on the upstream dam face may be calculated by the following formula:

$$\tau^s = (p + \overline{p}_y - \sigma_y^s)n \qquad (F.3.3)$$

Where:

τ^s = shear stress on the upstream dam face, MPa;

p = intensity of water pressure that the calculation section bears on the upstream dam face (sediment pressure shall be included if any), MPa;

$\overline{p}_y$ = intensity of uplift pressure that the calculation section bears on the upstream dam face, in MPa;

n = slope of upstream dam face.

F.3.4 The shear stress on the downstream dam face may be calculated by the following formula:

$$\tau^{xi} = (\sigma_y^{xi} - p' + \overline{p}_y')m \tag{F.3.4}$$

Where:

τ^{xi} = shear stress on the downstream dam face, MPa;

σ_y^{xi} = vertical normal stress on the downstream dam face, MPa;

$\overline{p}_y'$ = intensity of uplift pressure that the calculation section bears on the downstream dam face, MPa;

m = slope of downstream dam face.

F.3.5 The horizontal normal stress on the upstream dam face may be calculated by the following formula:

$$\sigma_x^s = p + \overline{p}_y - (p + \overline{p}_y - \sigma_y^s)n^2 \tag{F.3.5}$$

Where:

σ_x^s = horizontal normal stress on the upstream dam face, MPa.

F.3.6 The horizontal normal stress on the downstream dam face may be calculated by the following formula:

$$\sigma_x^{xi} = p' + \overline{p}_y' + (\sigma_y^{xi} - p' + \overline{p}_y')m^2 \tag{F.3.6}$$

Where:

σ_x^{xi} = horizontal normal stress on the downstream dam face, MPa.

F.3.7 The principal stresses on the upstream dam face may be calculated by the following formulas:

$$\sigma_{z1}^s = (1 + n^2)\sigma_y^s - n^2(p + \overline{p}_y) \tag{F.3.7-1}$$

$$\sigma_{z2}^s = p + \overline{p}_y \tag{F.3.7-2}$$

Where:

σ_{z1}^s = the first principal stress on the upstream dam face, MPa;

σ_{z2}^s = the second principal stress on the upstream dam face, MPa.

F.3.8 The principal stresses on the downstream dam face may be calculated by to the following formulas:

$$\sigma_{z1}^{xi} = (1 + m^2)\sigma_y^{xi} - m^2(p' - \overline{p}_y') \tag{F.3.8-1}$$

$$\sigma_{z2}^{xi} = p' - \overline{p}_y' \tag{F.3.8-2}$$

Where:

σ_{z1}^{xi} = the first principal stress on the downstream dam face, MPa;

σ_{z2}^{xi} = the second principal stress on the downstream dam face, MPa.

The calculation formulas presented above for vertical normal stresses and others, on the upstream and downstream dam faces, are suitable for the situations without uplift pressures. When uplift pressures exist, the formulas of τ^s, τ^{xi}, σ_x^s, σ_x^{xi}, σ_{z1}^s, σ_{z2}^s, σ_{z1}^{xi}, and σ_{z2}^{xi} shall be in accordance with those for concrete gravity dams with uplift pressures taken into account, referring to SL 319—2005.

F.4 Calculation of Internal Stresses

F.4.1 The vertical normal stresses inside the dam may be calculated by the following formulas:

When $x < T - T_1$,

$$\sigma_y = a + bx \tag{F.4.1-1}$$

When $x \geqslant T - T_1$,

$$\sigma_{yc} = n' + a + bx \tag{F.4.1-2}$$

$$a = \frac{\sum W}{A} - \frac{e_2 \sum M}{J} \tag{F.4.1-3}$$

$$b = \frac{\sum M}{J} \tag{F.4.1-4}$$

Where:

x = horizontal coordinate (with origin at the centroid of the expanded section and left as positive direction), m;

n' = ratio of the elastic modulus of impervious concrete slab to that of stone masonry.

F.4.2 The shear stresses inside the dam may be calculated by the following formulas:

1 Shear stress of stone masonry:
$$\tau = a_1 + b_1 x + c_1 x^2 \quad \text{(F.4.2-1)}$$
2 Shear stress of impervious concrete slab:
$$a_1 = \tau^{xi} \quad \text{(F.4.2-2)}$$
$$b_1 = b_m + \frac{\partial a}{\partial y} - \gamma_h \quad \text{(F.4.2-3)}$$
$$c_1 = \frac{1}{2} \frac{\partial b}{\partial y} \quad \text{(F.4.2-4)}$$
$$\tau^c = a_{1c} + b_{1c} x + c_{1c} x^2 \quad \text{(F.4.2-5)}$$
$$a_{1c} = a_1 + (b_1 - b_{1c})(T - T_1) + (c_1 - c_{1c})(T - T_1)^2$$
$$+ (n' - 1)[a + b(T - T_1)]\cot\alpha \quad \text{(F.4.2-6)}$$
$$b_{1c} = n'\left(b_m + \frac{\partial a}{\partial y}\right) - \gamma_h \quad \text{(F.4.2-7)}$$
$$c_{1c} = \frac{1}{2} n' \frac{\partial b}{\partial y} \quad \text{(F.4.2-8)}$$

Where:

n' = ratio of the elastic modulus of impervious concrete slab to that of stone masonry;

α = angle between impervious concrete slab and the horizontal surface, (°).

F.4.3 The horizontal normal stresses inside the dam may be calculated by the following formulas:

1 Horizontal normal stress of masonry body:
$$\sigma_x = a_2 + b_2 x + c_2 x^2 + d_2 x^3 \quad \text{(F.4.3-1)}$$
$$a_2 = \sigma_x^{xi} \quad \text{(F.4.3-2)}$$
$$b_2 = b_1 m + \frac{\partial a_1}{\partial y} - \lambda \gamma_h \quad \text{(F.4.3-3)}$$
$$c_2 = c_1 m + \frac{1}{2} \frac{\partial b_1}{\partial y} \quad \text{(F.4.3-4)}$$
$$d_2 = \frac{1}{3} \frac{\partial c_1}{\partial y} \quad \text{(F.4.3-5)}$$

2 Horizontal normal stress of impervious concrete slab:

$$\sigma_x^c = a_{2c} + b_{2c}x + c_{2c}x^2 + d_{2c}x^3 \qquad \text{(F. 4. 3 - 6)}$$

$$\begin{aligned}a_{2c} = {} & a_2 + (a_{1c} - a_1)\cot\alpha + [(b_{1c} - b_1)\cot\alpha - (b_{2c} - b_2)](T - T_1) \\ & + [(c_{1c} - c_1)\cot\alpha - (c_{2c} - c_2)](T - T_1)^2 \\ & - (d_{2c} - d_2)(T - T_1)^3 \qquad \text{(F. 4. 3 - 7)}\end{aligned}$$

$$b_{2c} = b_{1c}m + \frac{\partial a_{1c}}{\partial y} - \lambda \gamma_h \qquad \text{(F. 4. 3 - 8)}$$

$$c_{2c} = c_{1c}m + \frac{1}{2}\frac{\partial b_{1c}}{\partial y} \qquad \text{(F. 4. 3 - 9)}$$

$$d_{2c} = \frac{1}{3}\frac{\partial c_{1c}}{\partial y} \qquad \text{(F. 4. 3 - 10)}$$

$\dfrac{\partial a}{\partial y}$, $\dfrac{\partial b}{\partial y}$, $\dfrac{\partial a_1}{\partial y}$, $\dfrac{\partial b_1}{\partial y}$, $\dfrac{\partial c_1}{\partial y}$, $\dfrac{\partial a_{1c}}{\partial y}$, $\dfrac{\partial b_{1c}}{\partial y}$ and $\dfrac{\partial c_{1c}}{\partial y}$ are change rates of a, b, a_1, b_1, c_1, a_{1c}, b_{1c} and c_{1c} with y. The partial derivatives involved may be calculated directly, or in terms of differences by taking another section not far away from the considered section and calculating a, b, a_1, b_1, c_1, a_{1c}, b_{1c} and c_{1c}, i. e.

$$\frac{\partial a}{\partial y} \approx \frac{\Delta a}{\Delta y},\ \frac{\partial a_1}{\partial y} \approx \frac{\Delta a_1}{\Delta y},\ \cdots,\ \frac{\partial c_{1c}}{\partial y} \approx \frac{\Delta c_{1c}}{\Delta y}.$$

The formulas presented above of horizontal normal stresses inside the dam are suitable for the situations without uplift pressures. When uplift pressures exist, a correction term may be added to them according to the method specified in SL 319—2005.

Annex G   Calculation of Stresses of Gravity Block and Thrust Block with the Material Mechanics Method

G.1   Calculation Instructions

G.1.1   When the height of gravity block is lower than that of stone masonry arch dam, its stiffness may be approximately considered the same as the stiffness of foundation, and its foundation is assumed to be rigid, i.e. the influence of foundation deformation is not considered in the stress calculation of gravity block.

G.1.2   In the stress calculation of gravity block, the action on the interface of gravity block and arch abutment, of the arch abutment force system transferred from the arch dam and of partial water pressure, and the action of the water pressure directly applying on the upstream dam face, shall be taken into account and computed through the superposition of bending stresses in two directions.

G.2   Stress Calculation of Gravity Block

The stress calculation diagram of gravity block is shown in Figure G.2. The vertical normal stresses of points 1, 2, 3, and 4 on the calculation section may be calculated by the following formulas:

$$\sigma_{z1} = \frac{\sum W}{A} + \frac{\sum M_x}{I_x} \frac{b}{2} - \frac{\sum M_y}{I_y} \frac{a}{2} \qquad (G.2-1)$$

$$\sigma_{z2} = \frac{\sum W}{A} - \frac{\sum M_x}{I_x} \frac{b}{2} - \frac{\sum M_y}{I_y} \frac{a}{2} \qquad (G.2-2)$$

$$\sigma_{z3} = \frac{\sum W}{A} + \frac{\sum M_x}{I_x} \frac{b}{2} - \frac{\sum M_y}{I_y} \frac{a}{2} \quad (G.2-3)$$

$$\sigma_{z4} = \frac{\sum W}{A} - \frac{\sum M_x}{I_x} \frac{b}{2} + \frac{\sum M_y}{I_y} \frac{a}{2} \quad (G.2-4)$$

$$\sum W = W + G - U \quad (G.2-5)$$

Where:

σ_{z1}, σ_{z2}, σ_{z3}, σ_{z4} = vertical normal stresses of points 1, 2, 3, 4 on the calculation section, kPa;

$\sum W$ = sum of the vertical forces acting on the calculation section, kN;

$\sum M_x$, $\sum M_y$ = components around x, y axes of the sum of the moment with respect to the section centroid of all loads on the calculation section, kN·m;

I_x, I_y = moments of inertia of the calculation section with respect to x, y axes, m^4;

a, b = length and width of the calculation section, m;

W = self weight of gravity block, kN;

G = vertical force from arch abutment, kN;

U = uplift pressure at the bottom of gravity block, kN.

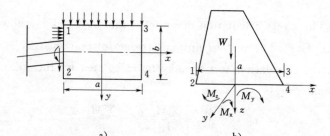

a)   b)

Figure G.2 Diagram of stress calculation of gravity block

a) Horizontal section view; b) Elevation view

The calculation formulas of the shear stresses, horizontal normal stresses and principal stresses on the upstream and downstream dam faces of the calculation section can be referred to SL 319—2005. In the calculation of the shear stresses at the bottom of gravity block, the torque M_z around vertical axis from arch abutment shall be considered.

As for the stress sign, compressive stress is positive normal stress. The positive directions of W, M_x, M_y are shown in Figure G. 2.

G. 3 Checking Calculation of Sliding Stability of Gravity Block

The stability against sliding of gravity block may be checked by the following formula:

$$K_2 = \frac{f\sum W}{\sqrt{H^2 + (V+P)^2}} \qquad (G.3)$$

Where:
H = normal force of arch abutment, kN;
V = shear force of arch abutment, kN;
P = water pressure acting on gravity block, kN.

The value of K_2 calculated by Formula (G. 3) shall not be less than that specified in Table 6. 3. 4.

G. 4 Stress Calculation of Thrust Block

The stresses of thrust block may be calculated by the formulas for stress calculation of gravity block presented in Annex G. 2, but in the terms containing $\sum W$, $\sum M_x$, and $\sum M_y$ in the formulas, the constraint effect of bedrock on the sides of thrust block shall be considered. The system of forces, including bedrock pressure, shear, moment, and seepage pressure, acting on the sides of thrust block transferred from bedrock shall be included, and some assumptions for simplification shall be made.

Explanation of Wording

Words used in this specification shall comply with the stipulations specified in the following table.

Clarification about wording used in this specification

Words in this specification	Equivalent expressions in specific situations	Strictness of requirement
Shall	It is necessary Is required to/it is required that Has to Only ... is permitted	Required
Shall not	Is required to be not/is required that ... be not Is not allowed/permitted/acceptable/permissible Is not to be	
Should	It is recommended It is advisable	Recommended
Should not	It is not recommended It is not advisable	
May	Is permitted Is allowed Is permissible	Permitted
Need not	It is not required that/no ... is required It is unnecessary	
Note: Except in certain cases (technical contents directly related to the construction quality, safety, hygiene, environmental protection and other public interests and requiring enforcement), generally do not use "must" and "must not".		